NOORTMAN & BROD

18th and 19th Century British Paintings

New York:
April 12—May 27, 1983

1020 Madison Avenue,
New York,
N.Y. 10021
Tel: (212) 772 3370
Telex: 968597

Weekdays: 9:30 – 5:30
Saturdays: 10:00 – 5:30

London:
June 14—July 29, 1983

24 St. James's Street,
London SW1A 1HA
Tel: 01-839 3871
Telex: 915570

Weekdays: 9:30 – 5:30
Saturdays: by appointment

All pictures are for sale and prices may
be obtained on application.

Introduction

We opened our gallery in New York in September 1981 and new premises in Maastricht, Holland, in May 1982. Thus, together with the marriage of the two galleries already in London, we have extended not only the geographical confines of our operation, but, also, the traditional fields of activity in which Robert Noortman and Thomas Brod have previously been active. I am grateful, therefore, to them, for having been given the opportunity to organise this present exhibition which we hold to coincide specifically with the celebrations taking place in New York during April and May for "Britain Salutes New York" and which will be our summer exhibition at the St. James's Street gallery. It will provide an opportunity to see paintings covering two centuries of British art and to make observations about various aspects of the development of social and political patterns in England during these years.

Rodney Merrington

1

Francis Cotes, R.A.

1726-1770

A Boy in a Green Coat

Canvas: 77 × 64 cm.
          30 × 25 in.

Painted circa 1764

Collections: Eggan and Co., Farnham, Surrey
               Arthur Moss, Painswick, Gloucestershire
               Private collection, U.S.A.

Exhibited: London, Thos. Agnew & Sons Ltd., "Realism and Romance in
               English Painting", 1966, no. 27
               Nottingham, Nottingham University Art Gallery, "Introducing
               Francis Cotes", 1971, no. 23, pl. 13

Literature: E. M. Johnson, *Francis Cotes,* p. 76, no. 167

Francis Cotes was an able and successful portrait painter. He was particularly adept at painting the flawless complexions of both women and children. He began his career by working mainly in pastels, but from the 1750's onwards, he painted almost entirely in oils.

Our picture is painted in his characteristically fluent style, and possesses the clarity and charm which singled him out to be second only in sucession to Reynolds and Romney in his day.

There has been a revival of interest in Francis Cotes, and the Tate Gallery (London) has recently acquired a major picture by his hand.

Richard Dadd

1819-1887

Still-Life with Bottles and Corkscrew

Board: 26 × 20.3 cm.
　　　　　10¼ × 8 in.

Signed and dated 183–

Collections: John Rickett
　　　　　　　Private collection, Great Britain

Exhibited: London, Tate Gallery, "The Late Richard Dadd", 1974, no. 23
　　　　　　Edinburgh, Scottish National Portrait Gallery, December 5 1980—
　　　　　　January 31, 1981

Literature: Patricia Allderidge, *Richard Dadd,* 1974, p. 34, no. 18, illus.

The still-life has never featured prominently in the subject matter chosen by English artists, and it is certainly unusual for Richard Dadd who is principally remembered for his fantasy pictures of fairies and other "wee folk". This is an early work for the artist, painted while he was still studying in the traditional academic manner. It is certainly reminiscent of the European still-life tradition, and can also be compared to those watercolours of similar subject matter by Peter de Wint. Patricia Allderidge has pointed out that if the painting dates from 1834, as has been suggested, then it is the earliest known painting by Dadd to survive, dating from the year in which his family moved from Chatham to London. However, she feels that a slightly later date is more likely.

It is interesting to note that the crumpled handkerchief of this composition is recalled in the later *Portrait of a Young Man* painted in Bethlehem Hospital in 1853, to which institution Dadd had been confined ten years earlier for murdering his father. Dadd never recovered from his madness and spent the considerable remainder of his life in either Bedlam or Broadmoor Hospitals in London and Berkshire.

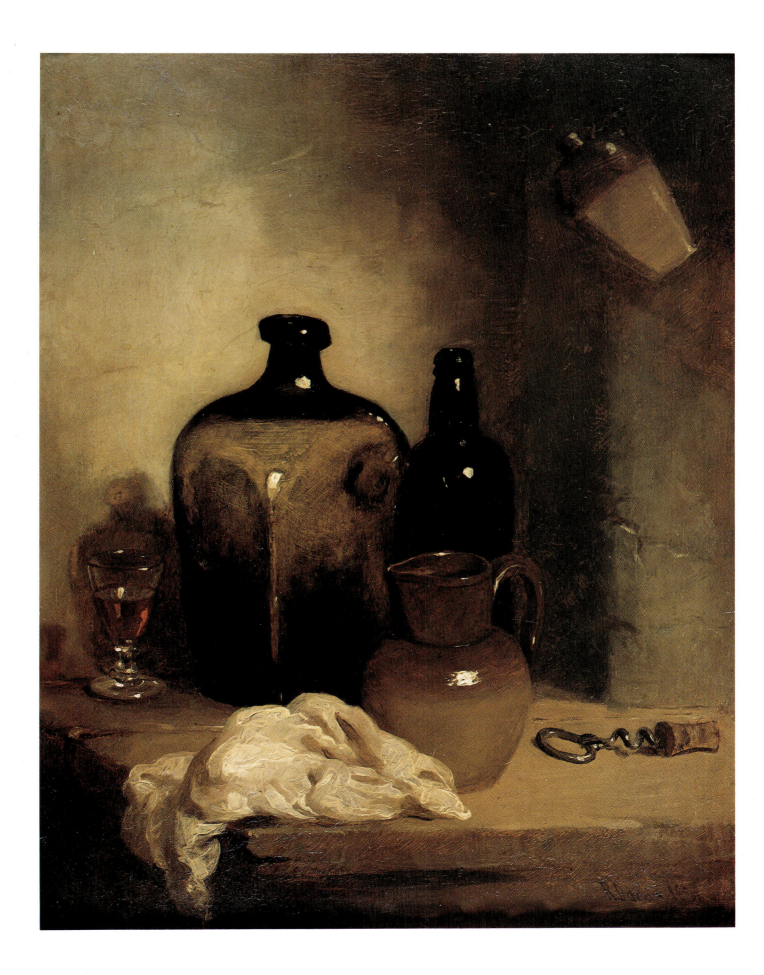

3

Thomas Gainsborough, R.A.

1727-1788

River Scene with Cattle Watering and Ferry Boat

Canvas: 124 × 99 cm.
          49 × 39 in.

Painted circa 1754–6

Collections: Painted for Thomas Spencer, Hart Hall, Suffolk
               Lord Knaresborough (Sir H. Meysey-Thompson) into whose family
               the picture probably came with Mary Spencer of Hart Hall in 1769
               Private collection, U.S.A., 1976
               Private collection, Great Britain

Exhibited: London, Tate Gallery, "Thomas Gainsborough", 1980–81, no. 86
               Paris, Grand Palais, "Gainsborough", 1981, no. 21

Literature: Jeffery Daniels, "Gainsborough the European", *The Connoisseur*
               (review of the exhibition in Paris), February 1981, vol. 206, no. 828,
               p. 111, illus.
               John Hayes, *Thomas Gainsborough* (exhibition catalogue for the Tate
               Gallery), 1980, p. 21, p. 98, no. 86, illus. p. 97 & in colour on back of
               cover
               John Hayes, *Gainsborough's Landscape Paintings,* 1982, no. 56
               Ellis Waterhouse, *Gainsborough,* 1958, p. 107, no. 827 and illus. pl. 37

The major retrospective exhibition of works by Gainsborough, both paintings, watercolours and
drawings, organized by John Hayes in London and Paris during the Winter of 1980/81 has enabled us
to take a fresh look at this great master of the eighteenth century in the context of European painting
as a whole. The influences not only of the seventeenth century Dutch masters such as Cuyp, Jacob
Ruisdael, Van der Heyden and Wynants, but also of the French School, Watteau, in particular,
become very evident. In Hubert Gravelot, Gainsborough had a distiguished French artist as a teacher.
The very essence of the artist also sets him somewhat apart from his contemporaries. He was
gregarious and a brilliant conversationalist, passionately fond of music and the theatre, a splendid
letter writer and these qualities combined to bring an effervescence to his paintings which has
outstripped his rivals. He possessed a lightness of touch and a great mastery of perception especially
in his portraiture.

continued on following page

[3 Thomas Gainsborough, R.A.]

In this landscape with a river view, we have an example of the artist's relatively early work. It must have been commissioned specifically, by Thomas Spencer of Hart Hall, as an over-mantel, and can be compared with two other upright paintings commissioned for Woburn Abbey by the Duke of Bedford.

John Hayes in his excellent catalogue for the Tate exhibition, describes it thus:

> Warm in tonality, and rococo in its rhythmic composition, the handling of the pinks and yellows at the horizon and the glow which permeates the whole canvas seem to have been inspired by Cuyp, whose work was then beginning to become popular among British collectors. (John Hayes, *Thomas Gainsborough,* London: The Tate Gallery, 1980, p. 98, no. 86).

Gainsborough was never content merely to mirror nature, indeed the conventions and tastes of his day were not sympathetic to this point of view. One is, however, tempted to wish that he had painted more pictures like the National Gallery's *Mr. and Mrs. Andrews in a Park,* surely one of the freshest masterpieces of the century. His art was a synthesis of those aspects of nature which he considered important, and it is perhaps appropriate to quote, here, his opinion expressed to a patron who had asked him to paint a specific topographical view. Gainsborough wrote to Lord Hardwicke thus:

> Mr. Gainsborough presents his humble respects to Lord Hardwicke, and shall always think it an honour to be employed in anything for his Lordship, but with respect to real views from Nature in this country he has never seen any place that affords a subject equal to the poorest imitations of Gaspar or Claude. Paul Sandby he believes is the only man of genius who has employed his pencil in that way. Mr. G. hopes that Lord Hardwicke will not mistake his meaning, but if his Lordship wishes to have anything tolerable of the name of Gainsborough, the subject altogether, as well as figures &c., must be of his own brain; otherwise Lord Hardwicke will only pay for encouraging a man out of his way, and had much better buy a picture of some of the good Old Masters. (Ellis Waterhouse, *Gainsborough,* 1958, p. 15).

This is undoubtedly one of the finest landscapes to have left his easel and we are happy to be able to include it in this exhibition.

[detail]

Thomas Gainsborough, R.A.
1727-1788

Portrait of H.R.H. William Henry, Duke of Gloucester, K.G. (1743-1805), in the Uniform of Colonel of the 1st Regiment of Foot Guards, later the Grenadier Guards, Wearing the Ribbon and Star of the Garter

Canvas: 232.4 × 140.3 cm.
            91¼ × 55⅛ in.

Painted circa 1775

Collections: Mrs. Gainsborough's sale, Christie's, London, April 4, 1797, no. 35
Bought by Hammond
Princess Sophia Matilda of Gloucester (gift of the Prince Regent, July 9, 1816)
Earl Waldegrave sale, Christie's, London, June 27, 1958, no. 57 illus.
Bought by Leggatt Bros.
Private collection, Great Britain

Exhibited: London, Royal Academy, "Winter Exhibition: Works by Old Masters and by Deceased Masters of the British School", 1899, no. 150

Literature: Sir Walter Armstrong, *Gainsborough*, 1899, p. 196
Violet Biddulph, *The Three Ladies Waldegrave*, 1938, p. 94 illus.
Oliver Millar, *The Later Georgian Pictures in the Collection of Her Majesty the Queen*, II, 1969, p. 35
Ellis Waterhouse, *Gainsborough*, 1958, p. 71, no. 317
Ellis Waterhouse, *Preliminary Check List of Portraits by Thomas Gainsborough*, Walpole Society, XXXIII, 1953, p. 49
William T. Whitley, *Gainsborough*, 1915, p. 116

William Henry was the third son of Frederick Louis, Prince of Wales (1707-1751), son of George II (1683-1760). He was born on November 14, 1743, created Duke of Gloucester and Edinburgh and Earl of Connaught on November 19, 1764, and married Maria, Countess Dowager Waldegrave on September 6, 1766. He was promoted in 1772 from Lieutenant-General to General, and later appointed a Field-Marshal in 1793. He had been Colonel of the 1st Regiment of Foot Guards since 1770.

According to Waterhouse, this picture is unfinished and was perhaps begun in 1775. If this assumption is correct, Gainsborough had, therefore, arrived to take up residence in London in the preceding year. He had quickly established a good clientele and was so busy that he did not actually exhibit at the Royal Academy during 1773 to 1775. This portrait was probably commissioned by the Duke's 1st Regiment of the Foot Guards, as opposed to the Grenadier Guards, and history does not relate why it was never finished or delivered. The painting of the uniform and the beautifully sketchy rendition of the landscape show the easy mastery of Gainsborough's later period.

5

Sawrey Gilpin, R.A.

1733-1807

Mare and Stallion in a Landscape

Canvas: 63.5 × 76.2 cm.
                25 × 30 in.

Collection: Private collection, Great Britain

Gilpin had the good fortune not only to be apprenticed to Samuel Scott, who had taken on the English mantle of Canaletto after the latter's visit to Britain, but, also, to have secured as patron the Duke of Cumberland. It was while in the Duke's employ that Gilpin became interested in painting horses. His Grace owned an important stud, thus giving the artist ample opportunity to develop his interest in sporting subjects. He is at his best while portraying animals in a landscape, and in this picture we have a perfect example of the harmonious facility which he was able to bring to such a subject.

6

William Havell

1782-1857

View of Lake Garda looking across to Salo, Italy

Canvas: 43.5 × 66 cm.
                17 × 26 in.

Collections: Lord Adam
                    Sir James Adam, Colne Park, Essex

In 1804, the "Old Society of Painters in Watercolour" was founded by William Havell and his friends such as Joshua Cristall and the Varley brothers whom he had met on sketching tours in Wales. Havell maintained a successful watercolour practice for much of his life, painting scenes in his local Thames Valley, occasionally in the Lake District, and in more exotic locales such as China and India where he was a portrait painter for eight years. It was after his trip to Italy in 1829-30 that Havell established himself as a painter in oil, virtually ceasing to paint in watercolour altogether. The picture of Lake Garda, with the sun's reflection on the water, recalls the work of Turner, to whom it has been previously attributed, and Havell may perhaps have had the master's Italian and Swiss watercolours in mind when he painted it.

7

William Hogarth

1697-1764

Southwark Fair

Canvas: 120.5 × 151 cm.
         47½ × 59½ in.

Signed and dated 1733

Collections: Mary Edwards, 1746
              D. Cameron, 1796-7
              T. Johnes, 1807
              4th Duke of Newcastle, before 1833
              Bequeathed by the 7th Duke of Newcastle to the Earl of Lincoln, 1937
              Sir Harry Oakes
              Thence by descent

Exhibited: London, British Institution, 1844, no. 140, no. 150
            Manchester, "Art Treasures", 1857, no. 31
            International Exhibition, 1862, no. 84
            London, British Institution, 1866, no. 127
            Nottingham, 1878
            London, Royal Academy, London, 1885, no. 144
            London, Royal Academy, London, 1908, no. 87
            London, Whitechapel Art Gallery, 1911
            Wembley, 1925
            New Brunswick, Maine, on loan to Bowdoin College
            London, Tate Gallery, "Hogarth Exhibition", 1972, no. 74
            Mexico City, Museo Nacional de Bellas Artes
            Milan, Palazzo Reale, "British Paintings 1660-1840", 1975, no. 27

Literature: R. B. Beckett, *Hogarth*, 1949, illus. pl. 61 & pl. 63, pp. 12, 73
             Austin Dobson, *William Hogarth*, 1891, pp. 57-58, 258
             Austin Dobson, *William Hogarth*, 1892, pp. 53, 171, 202
             S. Ireland, *Graphic Illustration of Hogarth*, I, 1974, pp. 110-111
             J. Nichols & G. Steevens, *The Genuine Works of William Hogarth*, I, 1808, pp. 84-94
             Ronald Paulson, *Hogarth's Graphic Works*, revised ed., I, 1970, pp. 154-8
             Ronald Paulson, *Hogarth: His Life, Art, and Times*, I, 1971, pp. 318-23

Hogarth's childhood and early life were beset by penury and unhappiness. His parents were educated folk but were forced into the confinement of the Rules of the Fleet Prison, as a result of the bankruptcy of their coffee shop, and for five years during their imprisonment, the young William suffered the deprivation resulting from this sad state of affairs.

Apprenticed to a silver engraver at the age of 17, he was forced to waste many valuable years at this trade. It was during this time, however, that he learned the methods of print-making. Had circumstances been more fortunate for him, he would have been able to develop his talents earlier.

continued on following page

His first success came with the publication of engravings from Samuel Butler's "Hudibras and the Lawyer", a mock, heroic poem. This was published in 1726. It was followed by the publication of *The Beggar's Opera* and *The Harlot's Progress*. Both these series were sold by subscription directly from the artist and earned him considerable financial reward.

In the early 1730's he painted numerous portrait commissions and conversation pieces, but was still missing the final accolade of Royal patronage. In this he was thwarted by William Kent and the Duke of Grafton who had reason to dislike the artist. They advised the Queen that it would be unsuitable for her to be painted by Hogarth, for she would be compromised by allowing her name to be linked with a painter who had reaped such rich rewards out of a subject so low and base, as *The Harlot's Progress*. Following on this disappointment he conceived his most famous series of pictures, *The Rake's Progress*, and our painting followed soon after. Hogarth received adequate compensation as the subscriptions to *The Rake's Progress* and *Southwark Fair* were quickly taken up.

Our painting can be dated after August 23, 1733, and was referred to by the artist in his advertisements for the print as "A Fair or The Humours of a Fair". Ronald Paulson (*Hogarth's Graphic Works,* revised ed., I, 1970, introduction and catalogue) suggests that Hogarth may well have been planning, if not actually working on the idea of *Southwark Fair* for some years. The artist drew on the memories of his youth, when he frequently visited the famous Saint Bartholomew's Fair. Hogarth spent the summer months near the Thames at Southwark, and it was there that he collected the abundant source of anecdotal subject matter which went into the conception of this picture.

It has been suggested that Hogarth purposely chose this bawdy, common subject in direct contrast to another subject engraved by a rival some time before, which depicted a Royal wedding, that of the Princess Royal and the Prince of Orange with all the extravagant panoply of the court.

Obviously, a great deal takes place in this animated composition. Paulson has separated it into three layers with teeming humanity in the foreground, actors and comedians playing out dreams and illusions on the stages in the middle ground, and in the background the church steeple and buildings giving way to the open countryside and sky.

There are four different theatrical entertainments in progress each marked by a banner. First, the one at the right shows a stage collapsing. This is a performance of the "Fall of Bajazet" and there is an obvious pun relating the disintegrating stage and the content of the drama. Secondly, the play being performed in the central booth is "The Siege of Troy", originally produced in 1707, revived in 1724, and revived again after the publication of Hogarth's print, thus the large banner depicting the famous "Wooden Horse". A third production shows Punch and Harlequin, Punch's horse picking Harlequin's pocket. Finally, the gabled stage at the left presents two small hand puppets advertising a marionette show.

The rope flyer, coming down from the steeple is Thomas Ridman, who broke his neck in 1740 descending from the steeple of St. Mary Friars, Shrewsbury. The other trapezist may have been Signor Volante. The tall man in the green jacket is Maximillian Christopher Muller, "near Eight Foot high, his Hand a Foot and his Finger Six Inches long" (Paulson, *Hogarth's Graphic Works,* revised ed. I, 1970, p. 157). The central figure is a drummeress advertising a show booth. A small painting of the drummeress exists and history relates that having saved her from ill-treatment by the master of the theatrical company, Hogarth became much enamoured of the girl. There is a peepshow at the left, and cardsharpers are at work on the right, with a monkey dressed as a gentleman next to a dwarf playing a bagpipe. Hogarth accents and develops these figures masterfully.

All these elements combine to give a clear depiction of Hogarth's humour, and the vibrance and delicacy of his brush are so typical of the best of eighteenth century painting. Hogarth is the Jan Steen of his era.

It is interesting to note that the beautiful painting of Mary Edwards in a red velvet dress by Hogarth which is now in the Frick Collection (New York) is, in fact, a portrait of the first owner of our picture.

Southwark Fair (engraving); Jan. 1733/4; 13½ × 17¹³⁄₁₆ in.

An advertisement of December 22, 1733 *(London Journal)* announces:

MR. HOGARTH being now engraving nine Copper-plates from Pictures of his own Painting, one of which represents the Humours of a Fair; the other eight, The Progress of a Rake; [offers] the Prints by Subscription on the following Terms: Each Subscription to be one Guinea and a half; half a Guinea to be paid at the Time of subscribing, for which a Receipt will be given on a new etched Print, describing a pleased Audience at a Theatre; and the other Payment of one Guinea on Delivery of all Prints when finished, which will be with all convenient Speed, and the Time publickly advertised. The Fair being already finished, will be delivered to the Subscribers on sight of the Receipt, on the 1st Day of January next, if required; or it may be subscribed for alone at 5s. The whole payment to be paid at the Time of Subscribing, Subscriptions will be taken in at Mr. Hogarth's the Golden Head in Leicester-Fields, where the pictures are to be seen. (Paulson, *Hogarth's Graphic Works,* revised ed., I, 1970, p. 154)

Sir Thomas Lawrence, P.R.A.
1769-1830

Portrait of Miss Julia Peel

Canvas: 142 × 111.5 cm.
          56 × 44 in.

Collections: Painted for Sir Robert Peel, Bt.
               Sale, June 6, 1907, lot 178
               Charles John Wertheimer
               Sir George A. Cooper, Bt., Hursley Park

Exhibited: London, Royal Academy, 1828, no. 77 (as the daughter of the Rt.
             Hon. William Peel)
             Berlin, Royal Academy of Art, "British Masters", 1908, no. 82
             Brighton, Royal Pavilion, Art Gallery and Museum, " 'Festival of
             Britain', Sir Thomas Lawrence, P.R.A.", 1951, no. 5
             London, Thos. Agnew & Sons Ltd., "Loan Exhibition of Pictures by
             Sir Thomas Lawrence, P.R.A.", 1951, no. 7

Literature: Sir Walter Armstrong, *Lawrence,* 1913, p. 157
              Kenneth Garlick, *Sir Thomas Lawrence,* 1954, p. 53, illust. pl. 103
              Kenneth Garlick, *A Catalogue of the Paintings, Drawings and Pastels of
              Sir Thomas Lawrence,* Walpole Society, XXXIX, 1964, pp. 159-60
              Lord Ronald Sutherland Gower, *Sir Thomas Lawrence,* 1900, p. 154
              Mrs. Jameson, *Companion to the Most Celebrated Galleries of Art in
              London,*1844, p. 374, no. 91
              G. Peel, ed., *The Private Letters of Sir Robert Peel,* 1920, p. 82
              *A Catalogue of the Pictures by Old Masters of the English School and Works
              of Art forming the Collection of Sir George A. Cooper, Bt.,* 1912, p. 39
              "Catalogue Raisonné des Tableaux de la Collection de Sir Robert
              Peel", *Le Cabinet de l'Amateur et de l'Antiquaire,* IV, 1845, p. 74, no. 91

Engraved: Samuel Cousins, 1833

Lawrence produced some of the best portraits of children ever to be painted of which "Pinkie"
(Huntington Collection), Master Lambton, (private collection), and Julia Peel, must rate amongst his
masterpieces.

continued on following page

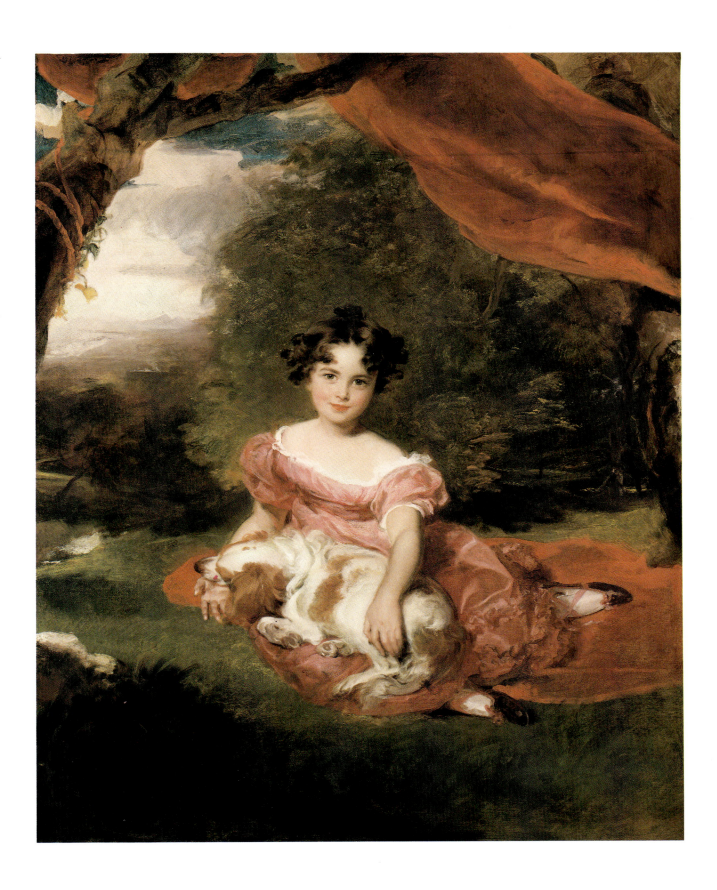

Julia Beatrice Peel, the eldest daughter of Sir Robert and Lady Peel, was born in 1821. She first married the Earl of Jersey in 1841 and secondly Charles Brandling of Middletown Hall in 1865. Lawrence had painted both her parents in the years preceding our picture. He painted her father twice, in 1825 and 1826, and the famous portrait of her mother Lady Peel, now in the Frick Collection (New York), was painted in 1828. This portrait was commissioned to hang as a pendant to Rubens's *Chapeau de Paille,* then in Sir Robert's collection.

It is of interest, to quote from a letter that Sir Robert wrote to his wife while staying with Lord and Lady Verulam at Gorhambury for a shooting weekend. The letter is dated January 1826:

> We have been out shooting, and had a very good sport. I think I killed about 36 head. I shall have a letter from you, my own love, tomorrow morning, and I do hope that it will bring me an excellent account of yourself and Bobby. I suppose that little Julia has had her second sitting today. Did she behave well, and is there any other time fixed by Sir Thomas for her to return to him? (G. Peel, ed., *The Private Letters of Sir Robert Peel,* 1920, p. 82)

The years 1825-30 see the culmination of Lawrence's career. He had, by this time, become a national hero, having painted the wonderful series of portraits now in the Waterloo Chamber at Windsor. He was President of the Royal Academy, honorary member of many academies in Europe, and also of the American Academy of Fine Art. He had achieved almost every honour known to an artist at that time. He advised the British Government on the successful acquisition of the Elgin marbles, and had encouraged the Angersteins to sell their collection to the nation at a lower price than could have been achieved elsewhere. He was also a trustee of the newly founded National Gallery

The engraving of 1833 by Samuel Cousins became a very popular print and was entitled *Childhood's Companion.* We exhibit a copy of the print.

Childhood's Companion (engraving), 1833

Sir Thomas Lawrence, P.R.A.

1769-1830

Portrait of John Allnutt (1773-1863)

Canvas: 236.2 × 144 cm.
           93 × 57 in.

Collections:	By descent to John Allnutt's granddaughter who married the 2nd Earl Brassey Commander Sidney Egerton, R.N. Captain Thomas Egerton
Exhibited:	London, Royal Academy, 1799, no. 5 Brighton, The Royal Pavilion, Art Gallery and Museum, " 'Festival of Britain', Sir Thomas Lawrence, P.R.A.", 1951, no. 7 London, Thos. Agnew & Sons, Ltd., "Loan Exhibition of Pictures by Sir Thomas Lawrence, P.R.A.", 1951, no. 1 London, National Portrait Gallery, "Sir Thomas Lawrence", 1979, no. 13
Literature:	Sir Walter Armstrong, *Lawrence*, 1913, p. 108 Kenneth Garlick, *Sir Thomas Lawrence*, 1954, p. 24, pl. 41 Kenneth Garlick, *A Catalogue of the Paintings, Drawings and Pastels of Sir Thomas Lawrence*, Walpole Society, XXXIX, 1964, p. 17 Lord Ronald Sutherland Gower, *Sir Thomas Lawrence*, 1900, p. 104

Painted in 1799, this portrait comes towards the end of a period in the painter's oeuvre during which he made a series of grand romantic pictures. These included *Satan Calling His Legions*, 1797, which belongs to the Royal Academy of Arts in London and the first of his three portraits of the famous actor, John Kemble, portrayed in theatrical roles.

On the death of Sir Joshua Reynolds in 1792, Lawrence accepted the post of "Painter in Ordinary" to the King, an honour partly due to his portrait of Queen Charlotte, dated 1790. This was quickly followed by his election to the Royal Academy in 1794 at the age of 25. Spurred on by these two honours, and after unsuccessfully tackling several large history pictures, he went on to produce some of his noblest full-lengths. Particularly famous are the portraits of Sarah Barrett, nicknamed "Pinkie" (Huntington Gallery, San Marino), Lady Louisa Manners, and the wonderfully inventive portrait of John, Lord Mounstuart, dressed as a Spanish grandee (1795). In these, as in our portrait of John Allnutt, the artist combines a theatrical pose with an expansive landscape and sky, a composition he was to use again in some of the famous portraits of the crowned heads of Europe after the defeat of Napoleon, now at Windsor Castle in the Waterloo Chamber.

John Allnutt was a generous patron of British artists, as well as a great collector. A wine merchant by trade, he remained a life-long friend of Lawrence. Indeed, at the time of the artist's death, Lawrence's estate owed Allnutt 5,000 pounds. Lawrence's other friends included the bankers Coutts and Angerstein, the latter's collection later forming the nucleus of the National Gallery.

The pose of the sitter and the horse relates to Reynold's *Captain Orme*, 1756, in the National Gallery (no. 681), though in reverse, and in an article in *The True Briton* (April 29), subsequently copied by *Lloyds Evening Post* (May 8-10), Lawrence was commended for avoiding his characteristic "wanton glitter" and for the portrait's "plain, sober, manly style of colouring, which we advise him to cultivate".

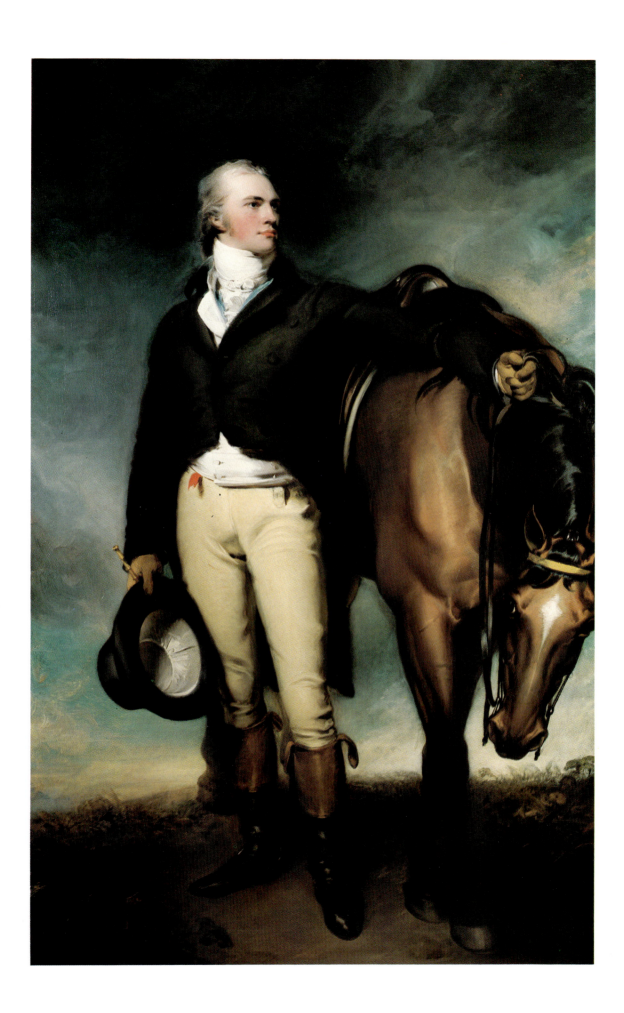

Sir Thomas Lawrence, P.R.A.
1769-1830

Marie Caroline, Duchesse de Berri (1798-1870)

Canvas: 91.5 × 73.7 cm.
          36 × 29 in.

Painted in 1825

Collections:	Marquis de Forbin, to whom given in 1830
	Comte de Marcellus
	John D. McIlhenny
	Mrs. John Wintersteen
	Private collection, Great Britain
Exhibited:	Paris, École des Beaux-Arts, "Portraits du Siècle", 1855, no. 185
	Worcester, Massachusetts, "Sir Thomas Lawrence", 1960, no. 19
Literature:	Sir Walter Armstrong, *Lawrence,* 1913, p. 114
	Kenneth Garlick, *Sir Thomas Lawrence,* 1954, p. 28, illus. p. 100
	Kenneth Garlick, *A Catalogue of the Paintings, Drawings and Pastels of Sir Thomas Lawrence,* Walpole Society, XXXIX, 1964, p. 35
	Lord Ronald Sutherland Gower, *Sir Thomas Lawrence,* 1900, p. 111, illus. p. 106
	G. S. Layard, *Sir Thomas Lawrence's Letterbag,* 1906, p. 199
Engraved:	J. Thompson, 1830

In August of 1825, Lawrence visited Paris to paint Charles X, a portrait commissioned by the English monarch, George IV. During the time Sir Thomas was working on this portrait, he met the famous duchess, who was very anxious to have herself painted by the renowned and popular English painter. Lawrence wrote back to George IV on September 4 for permission to make the portrait.

Marie Caroline de Bourbon was a famous society hostess whose elegance was renowned in court circles. She was the eldest daughter of Francis I of the Two Sicilies and married Charles, Duc de Berri, son of Charles X, in 1816. An anarchist assassinated her husband at the opera in 1820 at the time she was expecting their son, who became the Duc de Bordeaux and Comte de Chambord. It was through him that the duchess had pretensions to the throne of France, but her efforts came to nothing. In 1832, the Bourbon uprising was quelled and she, herself, was captured by the troops of Louis-Philippe.

There is a version of the picture at Versailles which may have been commissioned for another member of the family.

The duchess was also painted by Lefèvre and Baron Gerard, an unfinished portrait of whom Lawrence made in the same year.

Edward Lear
1812–1888

A View of Bethlehem and
A View of Mount Sinai (a pair)

Canvas: 24 × 47 cm. (each)
          9½ × 18½ in.

Both signed with monogram and titled, one inscribed and dated 1873 on the stretcher

Collection: Charles Allanson Knight, Esq.

Edward Lear showed a natural talent as a draughtsman at a tender age, receiving encouragement from one of his twenty elder brothers and sisters, Ann. His early natural history drawings earned him the patronage of the Earl of Derby in the mid 1830s, and it was for his children that Lear produced the famous *Book of Nonsense*. Indeed, Lear was such a success, both socially and artistically, that by 1846, he was giving Queen Victoria drawing lessons.

From 1837 onwards, partly for health reasons, but mainly due to a love of travel, Lear spent very little time in England. Besides extensively touring around the Mediterranean, he spent many months in the more exotic regions of the Middle East and India, currently much in vogue with the English traveller and artist.

Much of Lear's work was done in watercolour, and our pair of pictures of views of the Holy Land reflect Lear's highly personal style in the medium.

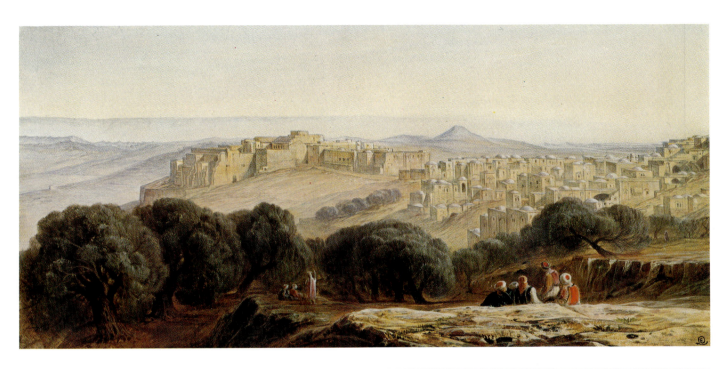

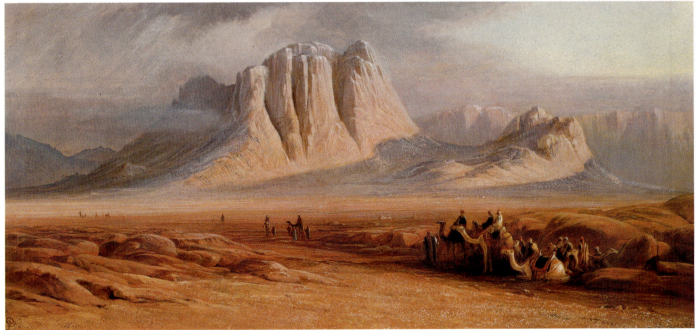

John Linnell

1792-1882

Lady Torrens and her Family

Canvas: 109.8 × 139.8 cm.
         43¼ × 55 in.

Signed and dated: I^NO. LINNELL: F: 1820

Collections: Sir Henry Torrens
              Private collection, U.S.A.

Exhibited: London, Royal Academy, 1821, no. 375
            London, Thos. Agnew & Sons Ltd., "The Portrait Surveyed: British Portraiture, 1670-1870", 1980, no. 44
            Cambridge, Fitzwilliam Museum, "John Linnell, A Centennial Exhibition", 5 October—12 December, 1982, no. 75
            New Haven, Connecticut, Yale Center for British Art, "John Linnell, A Centennial Exhibition", 26 January—20 March, 1983, no. 75

Literature: British Museum, Linnell's "Portrait Sketchbook"
             Katherine Crouan, *John Linnell, A Centennial Exhibition,* 1982, pp. 28-29, no. 75, illus. colour pl. no. 75
             Linnell MSS., "Journal" and "Autobiographical Notes"
             Story, *Linnell,* I, 115ff., II, p. 247

During his lifetime, John Linnell was held in high esteem as one of the most well-known artists of the period. He undertook several portrait commissions at the beginning of his career, of which *Lady Torrens and her Family* is outstanding.

Lady Torrens was the daughter of the English governor of St. Helena and met her husband when he was convalescing on the island while returning from a term of duty in Bombay. Torrens was knighted in 1814 and served as military secretary to the Duke of Wellington, and, by 1820, had been made Adjudant-General of the forces. By this time, Lady Torrens had borne her husband six children. Her friendship with Linnell continued for many years and she remained one of her most devoted admirers. John Varley, to whom Linnell had been apprenticed in 1804 at the tender age of 12, had evidently effected the introduction between Linnell and the Torrens'. Varley had many connections with rich patrons, some of whom had become his pupils in the art of painting in watercolours, a popular pastime for the gentry of the day. The success of this commission, therefore, was of prime importance to Linnell, especially at this point in his career. He exhibited the picture at the Royal Academy in 1821

Linnell had met Sir Thomas Lawrence in 1813, whose collection of Old Master drawings he may well have seen. His friendship with Lawrence began to develop about the time of the Torrens' portrait. He had been introduced to William Blake in 1818 who had taken him to the British Museum to study Old Master prints and had, also, been to see the distinguished collection of Old Master pictures belonging to Lord Suffolk.

The Torrens' portrait group, therefore, is deliberately conceived in a grand and formal manner, softened, however, by many anecdotal details such as the children's toys. The view behind the sitter from the terrace into a landscaped park reflects his recently acquired knowledge of early Flemish and German masters, and the manner in which the portrait is painted shows his interest and knowledge of Italian and particularly Venetian masters.

13

William Linton
1791–1876

A View of the Tiber and the Roman Campagna from Monte Mario

Oil on paper laid down on canvas: 23.8 × 34.3 cm.
                                     9⅜ × 13½ in.

Signed and dated 1829 and inscribed *Rome*

Linton spent much of his early career painting in the Lake District which was within easy reach of his birthplace, Liverpool. He also copied a number of pictures by Richard Wilson, and like him, went to Italy in 1828, returning there for a second visit in 1840 on his way to Sicily and Greece. It was his rendering of the classical landscape which earned him his reputation, and he successfully exhibited at the Royal Academy and the British Institution from 1817 to 1859. The influence of Wilson is clearly evident in our picture of the Italian countryside, painted while the artist was staying in Rome in 1829.

14

Ben Marshall

1767–1835

Priam—"Jockey up with Groom and Gentleman—Mr. William Chifney with his Colt, Priam. Sam Day Jr. up at Epsom."

Canvas: 70 × 90 cm.
          28 × 36 in.

Signed and dated 1830 and inscribed *Priam*

Collections: Cecil Brown
               Christie's sale, London, February 21, 1930, lot 106
               Bought by Knoedler, New York, 1,200 guineas
               Jack Dick, Greenwich, Connecticut

Literature: *Portraits of Celebrated Racehorses,* III, Taunton, p. 65

Marshall gained the reputation for being the most eminent sporting painter in the two decades following Stubbs and Gilpin. He rejected portraiture early on in his career, producing pictures for the *Sporting Magazine,* and learning much from the canvases of Sawrey Gilpin. He moved from London to Newmarket in 1812, a more likely place to acquire commissions, and remained there until his death in 1835. *Priam* illustrates Marshall's typical compositional device of delineating figures against an extensive horizon, in this instance with the colourful, crowded course and grandstand of Epsom making an effective backdrop to the victorious group in the foreground. Seldom has the atmosphere of the racing world been so brilliantly captured as in this superb picture of the 1830 Derby winner, who, two years later, went on to add the Goodward Cup to his trophies.

15

John Martin

1789-1854

Esau Selling His Birthright

Oil on paper laid down on canvas: 26.7 × 32.4 cm.
                                       10½ × 12¾ in.

Signed

Collection: The Lord Kinnaird

Engraved: Ebenezer Landells, pl. 16, in W. Westall and J. Martin's *Illustrations of the Bible* (1835)

John Martin was a painter of historical and biblical scenes, as well as of landscapes. Born in Northumberland, he moved to London in 1806 where he first worked as a painter of porcelain. By 1811, he had begun to exhibit at the Royal Academy which he continued to do until 1852. Martin was appointed "Historical Landscape Painter" to Princess Charlotte and Prince Leopold in 1816, but it was his painting of *Belshazzar's Feast,* 1820, and the subsequent engravings of it, that brought him to the public's attention. After this success, his work became increasingly fantastic and grandiose, often showing a considerable debt to Turner, as can be seen in his vast canvases in the Tate Gallery.

Our sparkling picture of *Esau Selling His Birthright* can be dated to the early 1830s. The subject was engraved by Ebenezer Landells for the *Illustrations of the Bible* in 1835. The book consists of 96 wood engravings of Old Testament subjects, 48 being after John Martin's sepia drawings, and 48 after William Westall. The Reverend Hobart Caunter wrote the individual descriptions and it was published by Edward Churton. The engravings were issued in twelve monthly parts of eight plates each, between March 1834 and March 1835. This was followed a year later by 48 illustrations of the New Testament, again with Martin and Westall sharing the task.

16

William James Muller

1812-1845

An Englishman Wearing Turkish Costume

Board: 41 × 26 cm.
         16¼ × 10¼ in.

Signed and dated 1839

Collection: Private collection, Great Britain

Muller was born in Bristol and first studied under J. B. Pyne as an apprentice. He quickly developed a taste for travel, being of Prussian parentage, and made frequent visits abroad. In the year our panel was painted, he is recorded to have made an extended tour of Greece, Egypt, Malta and Naples. His vivacious and confident style is perfectly represented in this charmingly strong study. Alas, the identity of the sitter has not come to light.

17

James Baker Pyne, R.B.A.
1800–1870

Whitstable Sands with Women Shrimping

Board: 61 × 81.3 cm.
          24 × 32 in.

Signed and dated 1840

Artist no. 271

Collection: Private collection, Great Britain

Pyne was self-taught and, as can be seen from the two paintings exhibited here, he was an admirer of Turner. Since Turner was the greatest British master of the nineteenth century, it is not surprising that he had such a singular influence on his contemporaries.

On the right-hand side of this composition, the construction standing up from the beach is one of the Martello Towers built along the English coastline as protection against the Napoleonic invasion.

18

James Baker Pyne, R.B.A.
1800–1870

Sunset on Whitstable Sands

Board: 57 × 78.8 cm.
          22½ × 31 in.

Signed and dated 1847

Artist no. 217

Collection: Private collection, Great Britain

A pendant to catalogue number 17, *Whitstable Sands with Women Shrimping,* this picture shows the beach from a slightly different angle and a view of the parish church seen from the distance.

Sir Joshua Reynolds, P.R.A.
1723-1792

Portrait of Mrs. Thomas Jelf Powis with her daughter, Catherine

Canvas: 231 × 144.7 cm.
          91 × 57 in.

Collections: The artist's studio sale, April 16, 1796
                Sold to Dodge
                The Earl of Denbigh
                Charles John Wertheimer
                Sir George A. Cooper, Bt.

Exhibited: London, British Institution, 1824, no. 136
                Birmingham, 1833, no. 42
                London, Royal Academy, 1873, no. 103
                London, New Gallery, "Guelph Exhibition", 1891, no. 152
                London, Royal Academy, 1894, no. 128
                Berlin, Royal Academy of Arts, 1908, no. 66
                Oxford, Ashmolean Museum, on loan 1941–45
                London, at Osterley Park, on loan

Literature: Sir Walter Armstrong, *Sir Joshua Reynolds,* 1900, p. 224
                Algernon Graves and William Vine Cronin, *A History of the Works of*
                Sir Joshua Reynolds, P.R.A., II, 1889, p. 766, illus., III, p. 880
                E. K. Waterhouse, *Reynolds,* 1941, p. 68
                A Catalogue of the Pictures by Old Masters of the English School and Works
                of Art Forming the Collection of Sir George A. Cooper, Bt., 1912, p. 27

The sitter married Thomas Jelf Powis of Berwick House, Shropshire, and her daughter, Catherine, married William, Viscount Fielding, son of the Earl of Denbigh, in 1791. Our picture was painted in June 1777.

The pictures which Sir Joshua painted during the decade of the 1770s show him demonstrating a strong debt to the Italian masters, whom he so admired. He was very preoccupied, also, at this time, with his responsibilities as the President of the newly founded Royal Academy and felt his views and theories to be much in the public eye.

The portrait of Mrs. Powis is formally posed in the grand manner and conceived, as were all the portraits of this period, with the idea that they were suitable to be exhibited in public.

His mastery of the textures of silks and furs is superbly shown here, as is his facility for the use of rich and harmonious colour.

20

George Romney R.A.
1734-1802

Portrait of Sir Henry Watkin Dashwood, 3rd Bt.

Canvas: 75 × 61.5 cm.
          29½ × 24¼ in.

Collections: Lady Anna Loftus
               Christie's sale, London, July 8, 1896, lot 45
               Bought by Tooth
               Dashwood family, Kirtlington Park
               David McCowan, Esq.

Literature: H. Ward and W. Roberts, *Romney,* II, 1904, p. 41

The sitter was born in 1745, the eldest son of Sir James Dashwood, a Member of Parliament for Oxford and builder of the mansion at Kirtlington. Sir Henry married in 1780, Mary Ellen, daughter of John Graham of the Supreme Court of Culcutta and Kinross, Scotland. He was a Member of Parliament for Woodstock, and died in June of 1828.

George Romney, R.A.
1734-1802

Portrait of Mrs. Henry Maxwell

Canvas: 238.6 × 147.2 cm.
            94 × 58 in.

Collections: Lee Priory sale, August 1834
                Bought by Mrs. Harrison Barham
                Champion Russell
                Charles John Wertheimer

Exhibited: London, Royal Academy, 1884, no. 197
                Berlin, Royal Academy of Arts, 1908, no. 95
                Oxford, Ashmolean Museum, on loan 1941-5
                London, Osterley Park, on loan

Literature: Arthur B. Chamberlin, *George Romney,* 1910, p. 119, 326-7
                H. Ward and W. Roberts, *Romney,* I, 1904, p. 102, II, pp. 92, 96,
                illus. opp. 94
                A Catalogue of the Pictures by Old Masters of the English School and Works
                of Art Forming the Collection of Sir George A. Cooper, Bt., 1912, p. 31

The sitter was the second daughter of Edward Brydges of Wootton Court, Kent. In 1760, she married Henry Maxwell of Eushot House, Hertfordshire.

Romney travelled two years in Italy, 1773-5, spending much time in Rome. He studied directly from the Antique and, also, from the great Italian masters, especially Raphael. During the years following his return to England, he produced some of his best works and they show a great debt to his years in Rome.

Painted only a few years after Reynolds' *Mrs. Jelf Powis and her Daughter,* the pose of this portrait is slightly more neoclassical, but we see the same accent on the noble posture predominantly used by the artists of the period. This elegant portrait with its flowing lines is a typical example of the linear grace and poise for which Romney was justly famed.

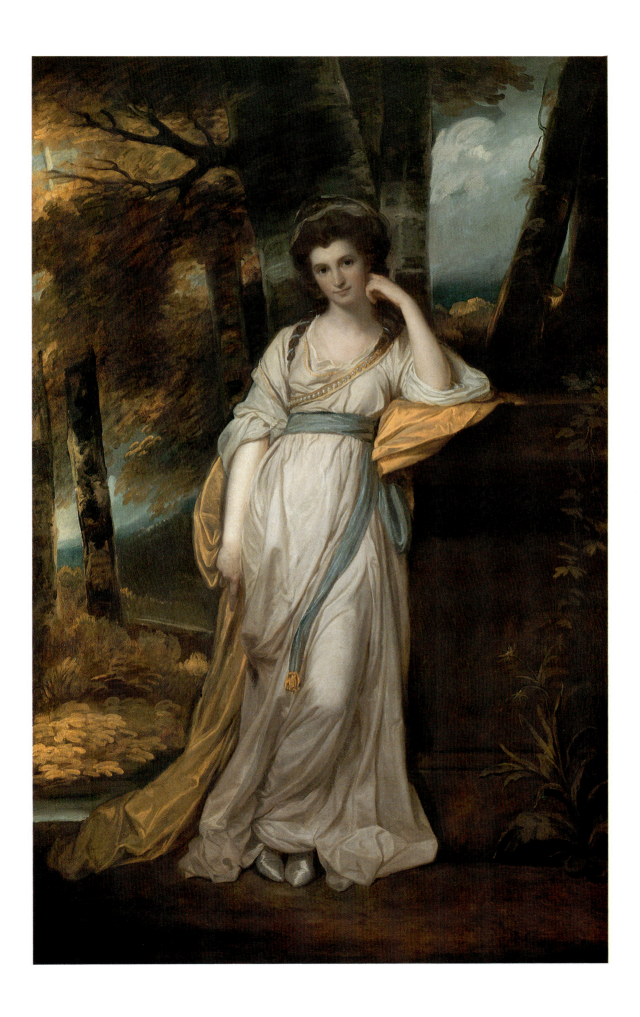

William Shayer, Sr.
1787-1879

Fishermen on the Beach

Canvas: 63.5 × 76.2 cm.
               25 × 30 in.

Collection: Private collection, Great Britain

William Shayer always idealised the plight of the English rustics in his canvases, and his treatment of simple country life and ways in a romantic manner explains the popularity his pictures have always enjoyed. Coastal scenes, such as this one, were a constant theme in his work, combining his skills of painting the people, animals and landscape of Victorian England.

23

James Stark

1794–1859

Milking Time

Panel: 45 × 60.5 cm.
          17¾ × 23¾ in.

Exhibited: London, Royal Academy, 1845

James Stark was a faithful follower of the founder of the Norwich School, John Crome, to whom he was apprenticed for three years in 1811. After Crome's death in 1821, it was primarily Stark who was praised by his contemporaries for adhering to his master's formula.

This picture was painted in the early 40s, after the artist had move to Windsor, never to return to his native Norwich. Our picture illustrates clearly Stark's predilection for the landscapes of Hobbema and Ruisdael, but it is generally lighter in touch and less sombre in mood than many of his earlier works.

24

Frederick W. Watts

1800-1862

View of Aylesford, Kent, from the River Medway

Board: 26 × 35 cm.
              10¼ × 14 in.

Collections: Private Collection, Great Britain

The work of Frederick Watts, a friend and follower of Constable, has undergone a much deserved revival over the past few years. This sparkling little picture shows the artist at his best, and he succeeds in describing the limpid light and serenity of an English summer day.

25

Frederick W. Watts

1800–1862

The Thames at Mortlake

Board: 25.3 × 35.3 cm.
              10 × 13⅞ in.

Inscribed on the reverse

The two pictures by Watts in this exhibition show the artist at the height of his powers, proving that he is at his best when painting on a small scale. It is interesting to make a comparison between this highly finished picture of the Thames and the more sketchy view of Aylesford (cat. no. 24), both of which capture similar atmospheric effects.

Sir David Wilkie, R.A.

1785-1841

The Duke of Wellington and his charger "Copenhagen"

Canvas: 59.7 × 47 cm.
          23½ × 18½ in.

Collections: The artist's sale, Christie's, London, June 30, 1842, lot 600
                Bought by Hogarth
                Sir Henry Russell, Swallowfield Park

Exhibited: Bath, "International Art Treasures Exhibition", 1973, no. 6

Literature: Lord Gerald Wellesley and John Steegman, *The Iconography of the First*
                Duke of Wellington, 1935, p. 51, no. 2

Wilkie exhibited a large portrait of the Duke of Wellington at the Royal Academy in 1834, and this now hangs in the Merchant Taylor's Hall, London. There are three other existing studies for the portrait, two of which are in the Wellington Collection.

The charger "Copenhagen" was the horse which the Duke rode at the Battle of Waterloo.

Wilkie's lively style and strong brushwork owe a great debt to some of the great Spanish masters and his visit to Spain after the war, 1817-18, affected his style in a very singular way. Although famous for his many captivating genre scenes of which *Chelsea Pensioners Reading the Waterloo Dispatch* (exhibited at the Royal Academy, 1822) was the most popular, his portraiture was very much appreciated towards the latter part of his career. It is interesting to compare the composition with that of *John Allnutt,* catalogue number 9.

Richard Wilson, R.A.

1713-1782

Roma from the Villa Madama

Canvas: 122 × 172.7 cm.
          48 × 68 in.

Collections: Leggatt, 1929
               Sold to Frederic Young
               By descent to Mrs. D. Doulton

Literature: W. G. Constable, *Richard Wilson,* 1953, p. 218, no. 1076

The inspiration for this grand classical view is obviously Claudian, with the composition coming more directly from Van Bloeman, called Orizzonte, who, in fact, painted the same view in 1736 (see David Solkin, *Richard Wilson,* Tate Gallery, cat. no. 67, p. 184)

The painting is made from the grounds of the Villa Madama which stands on the Eastern slope of Monte Mario. It shows the loggia, designed by Raphael for Pope Clement VII, on the right side of the composition with the Tiber flowing towards Rome and a distant view of the Alban Hills. We can just discern a glimpse of the sea behind the Castel San Angelo.

There are other versions of this composition, and the original commission probably came from the Earl of Dartmouth for whom Wilson painted a pendant entitled *Rome, St. Peters and the Vatican from the Janiculum* (Tate Gallery, London). Lord Dartmouth's view of Rome from the Villa Madama is dated 1753, and its pendant was probably made a little later. Our picture is not dated.

Ellis Waterhouse suggests that the role of Richard Wilson in relation to landscape painting in Britain in the eighteenth century is similar to that of Reynold's role in portraiture. He came from Wales, the son of a clergyman and was well educated. He received the patronage of a fellow countryman, Sir George Wynne, whose fortune was based on the discovery of a lead mine. Like many of his contemporaries, Wilson travelled to Italy in Autumn of 1750, and through the encouragement of Zuccarelli and, later, Claude Joseph Vernet, he broke away from portrait painting and concentrated on landscapes. In Rome, he met many English aristocrats making the Grand Tour, and their mutual enthusiasm for the Antique and Classical traditions formed Wilson's taste, and, on his return to Britain, he continued to paint in the same style that he had developed in Italy.

Richard Wilson, R.A.

1713-1782

'Cicero's Villa'

Canvas: 99 × 139.5 cm.
         37½ × 55 in.

Collections: This is almost certainly the painting referred to by Farington (see below) in 1810, so: Earl of Kerry
Mr. Hodges of Bath
William Wells, of Redleaf; his Sale, Christie's, May 20, 1852
Bought White
Spink & Co., 1931
Sir Robert Abdy, Bt.
Private collection, Great Britain

Exhibited: London, British Institution, "Old Masters", 1841, no. 122; 1843, no. 172: 'A grand Italian Landscape, with a ruined Temple, on a woody height above a river; a woman and child; and a man leading a horse in the foreground' (lent by William Wells)

Literature: W. G. Constable, *Richard Wilson,* 1953, no. 75a
Joseph Farington, *Diary,* X, 27 April 1810, p. 3643

Richard Wilson was as taken with the Italian landscape as were his patrons, who themselves followed in his footsteps in the Roman Campagna. This grand Italian landscape was known, probably even in Wilson's own time, with the title of 'Cicero's Villa', even though the building is a Roman tomb on the Via Nomentana that the artist sketched in his Italian Sketchbook of 1752 (now in the Victoria and Albert Museum, London). The setting is imaginary and evocative, much more so than the view of the same building included in the landscape by Joseph Farington after Wilson entitled 'In the Via Nomentana' (1776).

Farington is the first author to refer to the picture, which he does in his *Diary,* 27 April 1810:

> Wm. Wells I called upon and saw the large picture by Wilson formerly belonging to the Earl of Kerry; & lately to Mr Hodges of Bath. Mr Scroope bought it from Mr Hodges for 150 guineas & sold it to John Wells for Willm Wells for 200 guineas. Wm. Wells told me He had purchased the picture by Claude called 'The Inchanted Castle' from Mr Buchannan, the Picture dealer, for £1000, & that he should now cease from purchasing'.

It is interesting that William Wells should have so rounded off his collection with the finest of expressions of the Picturesque, by Claude Lorrain and Richard Wilson.

The slightly smaller version in the Manchester City Art Gallery has many differences in detail when compared with this work. It seems to have first come to light when it was lent to the 1877 Academy exhibition; after having belonged to the London dealer Vokins, it was bought from Agnew's by the gallery in 1897. When it was cleaned in 1949, it was discovered that the horse and man had been painted out.

Joseph Wright of Derby

1734-1797

A Moonlit Wooded Lake with a Castle

Canvas: 58 × 76 cm.
            23 × 30 in.

Signed with initials and dated 1788

Collections: Mr. Bird
                R.C.A. Palmer-Morewood
                Brighton, Brighton Art Gallery, on loan

Exhibition: Derby, England, 1934, no. 94

Literature: Benedict Nicolson, *Joseph Wright of Derby,* I, 1968, pp. 92, 269, no.
                333, II, colour pl. 292
                Ellis Waterhouse, *Painting in Britain 1530 to 1790,* 1953, p. 198,
                pl. 171a

Note: Mentioned in the artist's account book, no. 292, a companion
                moonlight, sold to Mr. Bird, Liverpool, £31.10.

Joseph Wright of Derby's early career was largely concerned with making pictures of scientific experiments, something unique for an English artist and which brought him considerable success. This was facilitated for him by his close connections with many of the industrial innovators in the Midlands at the time: Erasmus Darwin was a friend, he was in contact with the Lunar Society, and both Arkwright and Wedgwood were serious patrons. It was not until his extensive trip to Italy in 1774-5, where he witnessed Vesuvius erupting and the fireworks display in Rome at the Castel San Angelo, that he began to show an interest in landscape painting. Following his return home to his native Derbyshire, he commenced painting many local views especially in the vicinity of Matlock, and our picture is one of the most famous of these.

The play of light both artificial and natural was one of the most important factors in this artist's oeuvre. Moonlight was a constant source of fascination for Wright, and the beautiful rendering of it in this picture is characteristic of his achievement.

It is interesting to note that Professor Sir Ellis Waterhouse in the Pelican History of Art, *Painting in Britain, 1530 to 1790,* has chosen this picture to illustrate this facet of the artist's work.

Johan Joseph Zoffany, R.A.
1733–1810

The Plundering of the King's Cellar, Paris, 10th August 1792

Canvas: 99 × 127 cm.
        39 × 50 in.

Collections: Johann Zoffany sale, London, May 9, 1811, no. 94
              Anonymous sale, London, November 30, 1867, no. 88

Exhibited: London, Royal Academy, 1795, no. 18
            London, Royal Academy, "The First Hundred Years", 1951–2, no. 63
            London, National Portrait Gallery, "Johann Zoffany," 1977, no. 108
            reproduced

Literature: Lady V. Manners and G. C. Williamson, *John Zoffany*, 1920, p. 120–1
             Anthony Pasquin (John Williams), *An Authentic History of the Memoirs of the Royal Academicians,* 1796, p. 35–6
             Ellis Waterhouse, *Dictionary of British Eighteenth Century Painters,* Antique Collectors Club, 1981, p. 432

Engraved: Mezzotint by Richard Earlom, 1795

The title given to the picture when exhibited by Zoffany at the Royal Academy, 1795, was *Plundering the King's Cellar at Paris, August 10th 1793,* but he was presumably thinking of the sacking of the Tuilleries on August 10, 1792. The following in Joseph Farington's diary may possibly be connected with the present picture (August 1, 1794):

> Called on Zoffany . . . He was painting on of his
> Parisian subjects,—the woemen (sic) & sans
> culottes, dancing &c over the dead bodies of the
> Swiss soldiers. (J. Grieg, ed., *The Farington Diary,* 1922, p. 66)

Zoffany had visited Florence in 1772 where he stayed four years, and had subsequently made an extended visit to India in 1783–9. There he painted many portrait commissions and conversation pieces.

Our picture is fairly atypical of the artist's oeuvre and comes towards the end of his career. It is interesting to note the almost lugubrious fascination with the subject matter. When the picture was exhibited, it must have caused a great deal of interest, making, as it does, a lively commentary on a current event, and, also, embodying the Englishman's view of the social upheavals across the Channel.

A distinct parallel, here, can be drawn with our Hogarth, painted some sixty years earlier, and one is tempted to relate Zoffany's light and nervous brushwork to that of Hogarth.

Enquiries may be made to:—

NOORTMAN & BROD LTD

8 Bury Street,
St. James's,
London SW1Y 6AB
Telephone: 01-839 2606
Telex: 915570

NOORTMAN & BROD LTD

24 St. James's Street,
London SW1A 1HA
Tel: 01-839 3871

NOORTMAN & BROD LTD

1020 Madison Avenue,
New York,
N.Y. 10021
Tel: (212) 772 3370
Telex: 968597

NOORTMAN & BROD BV

Vrijthof 49,
6211 LE Maastricht
Holland
Tel: 043 16745
Telex: 56594

NOTES

Printed in England by: White Bros. (Printers) Ltd., 21-25 South Lambeth Road, London SW8 1SX   Telephone: 01-582 1282

JOSEPH CUNDALL
A VICTORIAN PUBLISHER

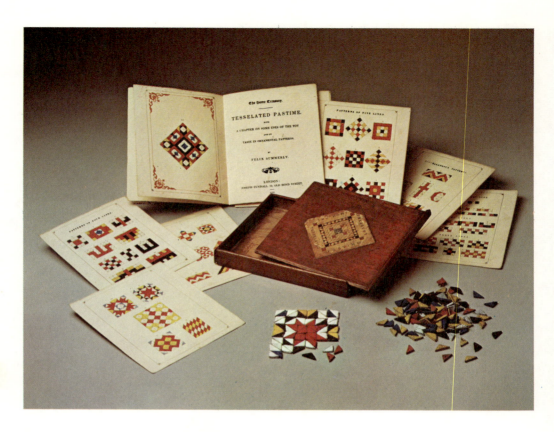

The Home Treasury *Tesselated Pastime*, 1843.
Collection Justin G. Schiller. Photograph by Don Henze, New York.

JOSEPH CUNDALL

A Victorian Publisher

Notes on his life and a check-list
of his books

by Ruari McLean

PINNER

Private Libraries Association

1976

DESIGNED AND PRODUCED FOR
THE PRIVATE LIBRARIES ASSOCIATION
BY RUARI MCLEAN ASSOCIATES LTD,
DOLLAR, SCOTLAND

© RUARI MCLEAN 1976

SBN 900002 13 1

PRINTED AND BOUND IN GREAT BRITAIN
BY T. & A. CONSTABLE LTD,
EDINBURGH

CONTENTS

Joseph Cundall's life

I. 12 Old Bond Street 1841-49 *p* 1

II. 21 Old Bond Street 1849-52 *p* 26

III. 168 New Bond Street 1852-68 *p* 33

IV. Last Years 1868-95 *p* 44

Towards a check-list of books published, edited or written
by Joseph Cundall

I. 12 Old Bond Street 1841-49 *p* 47

II. 21 Old Bond Street 1849-52 *p* 69

III. 168 New Bond Street 1852-68 *p* 75

IV. Last Years 1868-95 *p* 90

Index *p* 93

For Janna
great-granddaughter of Joseph Cundall
with affection

ACKNOWLEDGEMENTS

My search for information about Joseph Cundall led me to meet and enjoy the friendship of a very remarkable lady, Beatrice Wilford Cundall, widow of Rear-Admiral J. R. Cundall (a grand-son of the publisher), and her two children, Commander Rodney Cundall, RN, and Janna (Diana) Cundall, who is an architect.

Beatrice Cundall died in 1971, so my book is dedicated to her happy memory, and to her daughter Janna. I am infinitely grateful to them, and to various other members of the family, for the kindness, patience, hospitality and care with which they have met all my requests for help, loans, looking up things, etc, etc, during a period of over fifteen years.

I wish also to acknowledge much valuable help and courtesy from the Librarians and Staffs of the British Library, the Victoria and Albert Museum, the National Library of Scotland, the National Army Museum and the Royal Archive at Windsor Castle.

For both selling me and lending me precious books, I am grateful to many bookseller friends, especially Percy Muir and Laurie Deval, Max Brimmell, John Saumarez-Smith, Justin G. Schiller of New York, and Don and Edna Parkinson. For borrowings, help, advice, criticism, and friendship, I thank, especially, Fianach Lawry, Robin de Beaumont, John Porter and Nigel Temple. For patience, and the kindest of encouragement, I thank my wife.

Finally, I am grateful to the Private Libraries Association for undertaking to publish this book, and to David Chambers for exceptional patience and skilled help; and to the craftsmen of Constable's in Edinburgh for the trouble they have taken to try to print it as well as the subject deserves.

★   ★   ★

The check-list has been made with the sole purpose of identifying books on which Cundall worked. It is not a bibliography. If sufficient new information on Cundall's activities as a publisher is ever forthcoming, it might be desirable to compile a fuller bibliographical list.

I have not included descriptions of bindings (except in a few special cases) because of the fact that many of the books appeared in so many (often four or more) variants, for which nearly all documentary evidence is lacking. The only evidence we have is that of copies of the books that happen to have survived and been noticed.

As page sizes of all books are given in the checklist, I have not repeated this information in the captions beneath the illustrations, but merely indicated if a reduction has been made.

R. MCL.

Joseph Cundall, by an unknown photographer. Cundall family collection.

JOSEPH CUNDALL

I. 12 OLD BOND STREET 1841-49

family bible still in the possession of the Cundall family records that Joseph Cundall was born in Norwich on 22 September 1818. It is one of the comparatively few facts of his life that can be established from independent evidence: nearly everything else in this essay is deduced from the printed books which bear either his name as a publisher or his initials as editor. He was the second in a family of five: his father, Benjamin, was a substantial linen draper.[1] Both Joseph's grandfather and his great-grandfather were also called Joseph. According to *The Times*'s obituary, he was trained as a printer in Ipswich, and came to London in 1834, at the age of 16 (not, surely, having completed his training). There he worked for Charles Tilt, the well-known bookseller and publisher, whose shop front was the subject of a memorable etching by George Cruikshank.[2] By 1840, aged only 22, Joseph Cundall was himself an author, for Tilt published *Tales of the Kings of England: Stories of Camps and Battle-Fields, Wars and Victories, from the Old Historians*, by 'Stephen Percy', Cundall's earliest pseudonym. It is a pleasant little book of 240 pages, illustrated with wood-engravings after drawings by John Gilbert, an illustrator one year younger than Cundall, whose work appeared in many of Cundall's later books and who was knighted in 1872.

In 1841, Tilt published *Robin Hood and his Merry Forresters*, also by 'Stephen Percy', with illustrations again by Gilbert; the third edition of this book was published by Cundall himself in 1845.

In 1841,[3] Cundall, still only 23, became the successor to N. Hailes at the Juvenile Library, 12 Old Bond Street. Hailes had been the partner of John Sharpe (1777-1860), the proprietor of the Juvenile Library at the London Museum, 168 Piccadilly. In 1814, Sharpe & Hailes published *Sir Hornbook: or, Childe Launcelot's Expedition. A Grammatico-Allegorical Ballad*.[4] The text, in rhyme, was written by Thomas Love

---

[1] Benjamin Cundall's personal goods were worth between £3,000 and £4,000 when he died in 1875 (Norfolk and Norwich Record Office).

[2] In Tilt's *Comic Almanac*, March 1835, and shown in the writer's *George Cruikshank*, 1948.

[3] On 2 June 1841, Cundall opened an account with Coutts Bank with a deposit of £200. The account was closed in April 1844 for no known reason, and I have not been able to discover that he then opened an account elsewhere. The entries for his account in Messrs Coutts' ledgers record payments out to familiar names like Hullmandel, Leighton, Jane Leighton, Hailes (11 entries), Bogue, Tilt, H. Cole (first payment dated 3 May 1843, £22 1s.) and regularly to a Miss Crossthwaite; sources of payments in are not given. In Midsummer 1843 the account was balanced out at £1798 11s 4d. I thank Messrs Coutts for facilities to see these ledgers.

[4] Described and illustrated in *Early Children's Books and their Illustration*, Pierpont Morgan Library, New York, 1975. The illustrations in the Home Treasury edition were based on those in the 1814 edition.

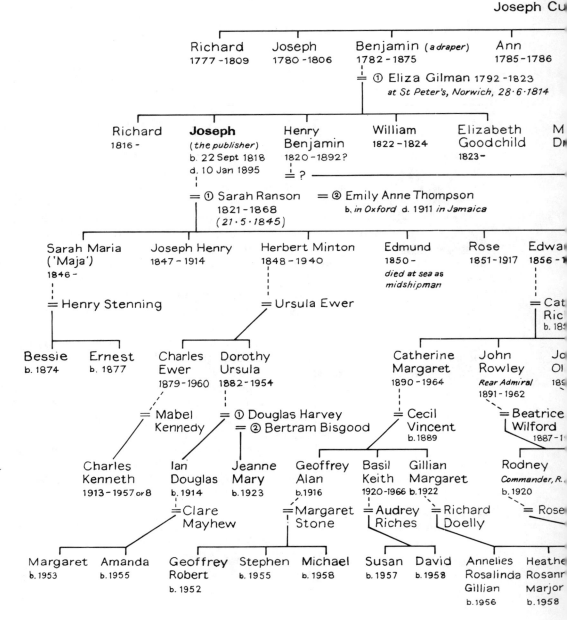

The Cundall family.

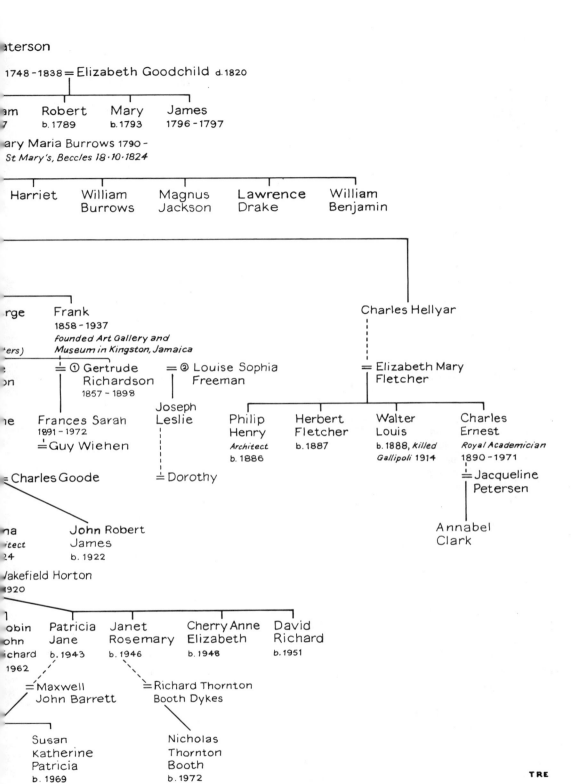

aterson

1748-1838 = Elizabeth Goodchild d.1820

	Robert	Mary	James
am	b.1789	b.1793	1796-1797

ary Maria Burrows 1790 –
St Mary's, Beccles 18·10·1824

Harriet	William Burrows	Magnus Jackson	Lawrence Drake	William Benjamin

rge   Frank
       1858-1937
founded Art Gallery and
Museum in Kingston, Jamaica

ers)

= ① Gertrude Richardson 1857-1898   = ② Louise Sophia Freeman

Charles Hellyar

= Elizabeth Mary Fletcher

Frances Sarah 1891-1972
= Guy Wiehen

Joseph Leslie

⋮ Dorothy

Philip Henry
Architect b.1886

Herbert Fletcher b.1887

Walter Louis b.1888, *killed Gallipoli* 1914

Charles Ernest
Royal Academician 1890-1971
= Jacqueline Petersen

= Charles Goode

na
itect
24

John Robert James b.1922

Annabel Clark

Jakefield Horton
920

obin ohn ichard 1962	Patricia Jane b.1943	Janet Rosemary b.1946	Cherry Anne Elizabeth b.1948	David Richard b.1951

= Maxwell John Barrett

= Richard Thornton Booth Dykes

Susan Katherine Patricia b.1969

Nicholas Thornton Booth b.1972

TRE

3

Peacock, who was a friend of Henry Cole (1808-82). How Cundall met Hailes, and what was the transaction by which he took over the Juvenile Library, is a mystery; but very soon afterwards, Henry Cole, then in his thirties and a rising Civil Servant, entrusted Joseph Cundall with an ambitious publishing project, called 'The Home Treasury', designed to reform children's books. The printer chosen for these books was Charles Whittingham the Younger, at the Chiswick Press, whose typography and press-work were at that time easily the best in Britain and possibly in the world. If the choice was Cundall's, he had developed early a discerning eye for printing.

Whittingham had also printed for Sharpe and Hailes; but Cole, who had been writing guides to places like Westminster Abbey and Hampton Court under the pen-name of Felix Summerly since 1833, did not as far as I know have a book of his own printed by Whittingham until *Handbook for the National Gallery*, published by Bell in 1843, the year in which the first Home Treasury books appeared. It looks as if Hailes introduced Cundall to Whittingham, and Cundall suggested him to Cole. Cole, as a father of young children and an energetic critic of everything he saw around him, including children's books, must have known the Juvenile Library and probably met Cundall through Hailes. His choice of Cundall was a remarkable tribute to the young man who had hardly begun his publishing career.

The manifesto for the Home Treasury, written by Cole, was as follows:

'The character of most Children's Books published during the last quarter of a century is fairly typified in the name of Peter Parley, which the writers of some hundreds of them have assumed. The books themselves have been addressed after a narrow fashion almost entirely to the cultivation of the understanding of children. The many tales sung or said from time immemorial, which appealed to the other, and certainly not less important elements of a little child's mind, its fancy, imagination, sympathies, affections, are almost all gone out of memory, and are scarcely to be obtained. The difficulty of procuring them is very great. Of our national nursery songs, some of them as old as our language, only a very common and inferior edition for children can be procured. Little Red Riding Hood and other fairy tales hallowed to children's use, are now turned into ribaldry as satires for men. As for the creation of a new fairy tale or touching ballad, such a thing is unheard of. That the influence of all this is hurtful to children the conductor of the proposed series firmly believes. He has practical experience of it every day in his own family, and he doubts not that there are many others who entertain the same opinions as himself. He purposes at least to give some evidence of his belief, and to produce a series of Works for children, the character of which may be briefly described as anti-Peter Parleyism.

'Some will be new Works, some new combinations of old materials, and some reprints carefully cleared of impurities, without deterioration to the points of the story. All will be illustrated, but not after the usual fashion of children's books, in which it seems to be assumed that the lowest kind of art is good enough to give the first impressions to a child. In the present series, though the statement may perhaps excite a smile, the illustrations will be selected from the works of Raffaelle, Titian, Hans Holbein, and other old masters. Some of the best modern Artists have kindly promised their aid in creating a taste for beauty in little children. All the illustrations will be coloured.

'In addition to the printed Works, some few Toys of a novel sort, calculated to promote the same object, will from time to time be published.'

Henry Cole was a skilful promoter and found it useful to disparage what was already on the market. Whatever the truth of his criticisms, and whatever the virtues of the texts he provided, it is certain that the Home Treasury series had the most

Cover design printed on coloured papers, for the Home Treasury series, after a design by Holbein for a binding in silver.
From *On Ornamental Art*, etc, by Joseph Cundall, 1848.

ULL. The Bull, as you see here, is stout and broad-chested, with a thick short neck, and a fine full eye. He lives in the fields among the Cows, which give us sweet milk, but the Bull gives us none; yet his flesh makes good food, called beef. He eats grass and herbs. His skin is thick and hairy, and of various colours, black, red, brown, and white; it is called a hide, and makes stout leather. He is not gentle like the Cow, but often

An Alphabet of Quadrupeds, Home Treasury, 1844. Reduced.

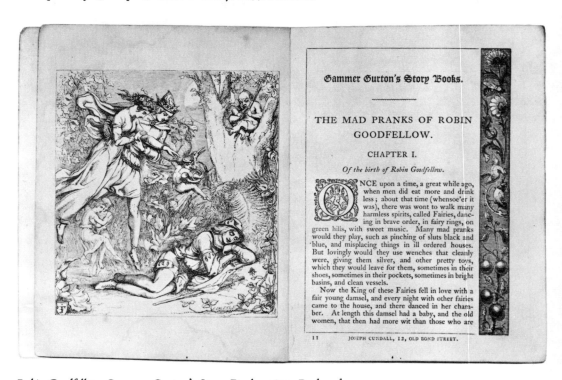

Robin Goodfellow, Gammer Gurton's Story Books, 1845. Reduced.

OUR and twenty tailors
   Went to kill a snail,
  The best man among them
   Durst not touch her tail.

She put out her horns
  Like a little Kyloe cow:
Run, tailors, run,
  Or she'll kill you all e'en now.

⁂

AY go up and gay go down,
   To ring the bells of London town.

  Oranges and lemons,
   Say the bells at St. Clement's.

Bull's eyes and targets,
Say the bells of St. Marg'ret's.

Brickbats and tiles,
Say the bells of St. Giles'.

Halfpence and farthings,
Say the bells of St. Martin's.

Pancakes and fritters,
Say the bells of St. Peter's,

Traditional Nursery Songs of England, Home Treasury, 1843.

Fruit! Oranges and Apples.   Buy my Straw-berries!

Any Chairs to mend?   Come and see the Giant!

Remember the Sweeper.   Pray think of Poor Jack.

Little Mary's Primer, 1847. A more visual approach to book design for children than in the Home Treasury: almost a strip.

Dust O! Dust O!

Who will buy my flowers?

Fish O!   All alive!

Do you want a link Sir?

Any knives to grind?

Who'll buy my images?

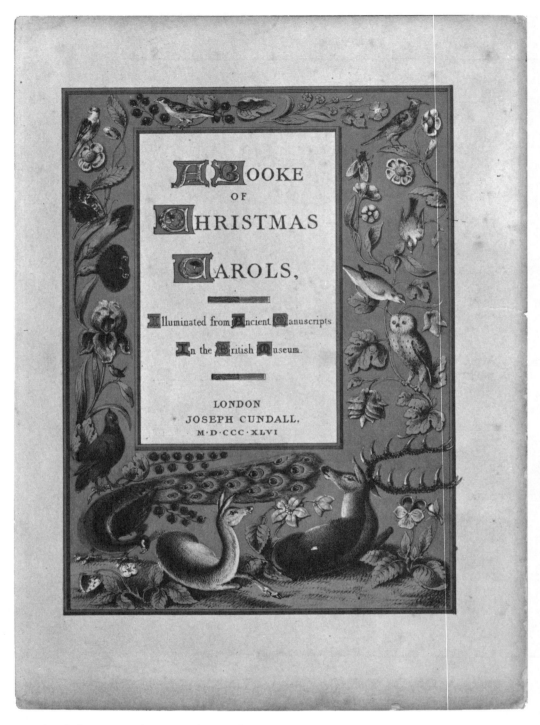

A Booke of Christmas Carols, 1846. Title-page, lithographed in colours and gold by Hanhart after drawing by John Brandard. *Facing*, the design for the title-page, drawn by John Brandard. From the original drawings in the possession of the Cundall family.

Dust O! Dust O!   Who will buy my flowers?

Fish O!   All alive!   Do you want a link Sir?

Any knives to grind?   Who'll buy my images?

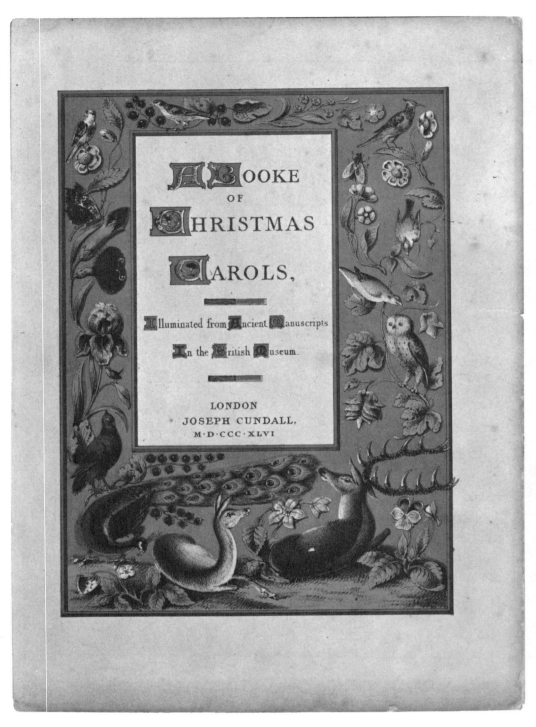

A Booke of Christmas Carols, 1846. Title-page, lithographed in colours and gold by Hanhart after drawing by John Brandard. *Facing*, the design for the title-page, drawn by John Brandard. From the original drawings in the possession of the Cundall family.

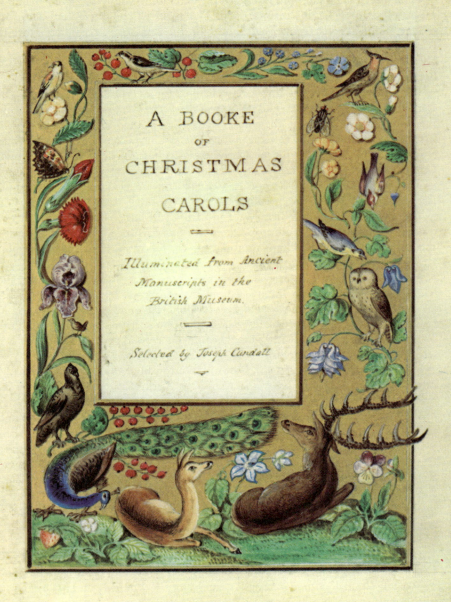

A BOOKE
OF
CHRISTMAS
CAROLS

Illuminated from Ancient
Manuscripts in the
British Museum.

Selected by Joseph Cundall

distinguished design treatment ever given to children's books up to this time. Like so many children's books of the period, they were basically adult in presentation: their covers, in paperback, were not pictorial but abstract Renaissance designs taken from Holbein and Dürer via Henry Shaw. The texts were meticulously printed in Whittingham's classically allusive typography, often in red and black, with pictures added. Cole did not make the extra jump ahead into story-telling by means of pictures, i.e. the cartoon strip. The artists he chose were either classical (Dürer, Holbein, even Teniers and Stoop) or 'the best modern' painters, from the Royal Academy: he did not use, for example, Cruikshank (then 51) or Tenniel or Doyle, who in 1843 were aged 22 and 19 respectively, but who had already both been published.

To our eyes most of the illustrations in the Home Treasury are charming but somewhat stilted. The chief innovation was that, after the first few titles, in which the colouring was by hand, the colour plates were printed; they were among the first children's books ever to have colour-*printed* illustrations. They were printed from wood-blocks, a few by the Chiswick Press but most by a new firm called Gregory, Collins & Reynolds, a group of young men who had all previously worked as apprentices to George Baxter. They were beautifully engraved and finely printed in clean, bright colours.

The printers in the new firm were probably still under the influence of their training by the idealistic Baxter, and their work was artistically better than most of the more commercial colour printing which followed later from Leighton, Kronheim and even Evans.

A further series in paper covers was then published, at sixpence each, called 'Gammer Gurton's Story Books', edited by 'Ambrose Merton, Gent., F.S.A.' (W. J. Thoms, 1803-85, editor of *A Collection of Early Prose Romances*, 1827-8, and founder in 1849 of *Notes & Queries*). These were also charmingly produced, printed by the Chiswick Press, with monochrome or colour-printed illustrations by Gregory, Collins & Reynolds.

It was the Home Treasury and Gammer Gurton books that Thackeray had in mind when he wrote, in *Fraser's magazine* for April 1846, 'let me be permitted humbly to move a vote of thanks to the meritorious Mr Cundall. The mere sight of the little books published by Mr Cundall, of which some thirty now lie upon my table, is as good as a nosegay.'

In 1843, another of Cole's commercial inspirations appeared: the first printed Christmas card. The design was commissioned from J. C. Horsley, later a knighted Royal Academician, and was printed lithographically and hand-coloured. George Buday in *The History of the Christmas Card*, 1954, quotes Cundall as saying that 'many copies were sold, but possibly not more than 1000',[1] at 1s each; and he reproduces the actual copy of the card sent (and dated 1843) by the artist to Henry Cole.

The imprint of the card is 'Published at Summerly's Home Treasury Office, 12 Old Bond Street, London'. Apparently no further Christmas cards were published by Cole or Cundall, although the idea caught on fairly quickly and before the end of the century was big business.

Cundall in those early years was a bookseller as well as a publisher, but no details of this side of his business have survived. At that time, many publishers kept

[1] In a letter written by Cundall published in *The Times*, 2 January 1884.

THE
LIFE AND ADVENTURES
OF
ROBINSON CRUSOE,

OF YORK, MARINER.

WITH AN ACCOUNT OF HIS TRAVELS

ROUND THREE PARTS OF

THE GLOBE.

London:

JOSEPH CUNDALL, 12, OLD BOND STREET;

THOMAS EDLIN, 37, NEW BOND STREET.

1845.

Robinson Crusoe, 1845, illustrated by Stothard.

bookshops, in which they sold other publishers' books as well as their own. (William Pickering had a most important antiquarian book business, and issued catalogues.) Cundall also ran a lending library for children, under the title of St George's Reading Library: an advertisement for it appeared in *Peter Parley's Annual* for 1848 (see p 60).

In 1844, Cundall began publishing 'Hazlitt's Holiday Library', edited by William Hazlitt: these were for children but produced in adult style, cloth-bound, the first two printed by the Chiswick Press, and each with four lithographed illustrations. Four titles appeared. Also in 1844, Cundall published the first of his 'art' books for an adult audience, *The Passion of Our Lord Jesus Christ*, edited by Henry Cole, and printed by the Chiswick Press (see p 15). This book was only the fourth edition of Dürer's thirty-seven woodcuts (known as 'The Little Passion on Wood' to distinguish them from 'The Little Passion on Copper'), first published at Nuremberg in 1511; and the illustrations were, remarkably, printed from stereotypes from the original wood-blocks, thirty-five of which had been sold to the British Museum in 1839. To the stereotype copies 'new borders were added; the worm-holes were cleverly stopped, and the injured portions were re-engraved with great care by that accomplished xylographer, Mr Charles Thurston Thompson. Mr Thompson also re-engraved the two missing subjects – No. 8, "Christ parting from his Mother", and the vignette on the title-page'.[1] Henry Cole, in his preface, claims that 'the process of stereotyping has had the good effect of restoring almost the original sharpness and crispness of the lines, and of rendering the present impressions nearer the state of the earliest impressions than they would have been had they been taken from the blocks themselves' – because of the slight shrinkage (sometimes as much as $\frac{1}{8}$ in. in a cast of 6 in.) of the plaster of paris used for making the cast from the original wood-blocks, into which the stereotype metal was poured. The Cole and Cundall Dürer was bound in leather, blind-stamped with a design adapted from a German binding of the fifteenth century, and was sold for £1 1s. Six copies were printed on vellum on one side of the leaf only: one of these, 'bound in purple velvet, with gilt clasp, monogram and corner ornaments in dead and bright gold, leather joints, and velvet inside borders, white watered silk fly-leaves and linings, gilt leaves, by Hayday, in a blue morocco case, with lock and key', belonged to Sir Joseph Walter King Eyton (Item 558 in the Catalogue of his Library Sale, 1848).

In 1845, the Royal Society of Arts, at the suggestion of Prince Albert, offered prizes for the design of a tea service and beer mugs, in an attempt to improve the appearance of domestic articles in common use. Henry Cole entered the competition enthusiastically, and enlisted the help of Herbert Minton, the china manufacturer, persuading him to design the beer mugs while he tackled the tea service. Both entries won silver medals and were exhibited in the summer of 1846. Their success fired Cole with ambitious plans for 'Summerly's Art Manufactures'. One thing now led quickly to another, culminating in the concept, planning, and triumphant opening of the Great Exhibition of 1851. From 1846 onwards, Cole had more interesting fish to fry than the Home Treasury of children's books: that had been a very considerable achievement, but it was now, as far as Cole was concerned, completed.

[1] Austin Dobson ed., *The Little Passion*, Bell, 1894.

The Passion of our Lord Jesus Christ,

pourtrayed by Albert Durer.

EDITED BY HENRY COLE,

AN ASSISTANT KEEPER OF THE
PUBLIC RECORDS.

London:

Joſeph Cundall, 12, Old Bond Street; William Pickering,
177, Piccadilly; George Bell, 186, Fleet Street;
J. H. Parker, Oxford; J. and J. J.
Deighton, Cambridge.

1844.

The Passion of Our Lord Jesus Christ, 1844.

Joseph Cundall had married,[1] on 21 May 1845, at Sproughton in Suffolk, Sarah Ranson of Sproughton, three years younger than himself; their first child, Sarah Maria (known in the family as 'Maja') was born at 12 Old Bond Street on 25 February 1846, and christened on 21 September at St George's, Hanover Square, her godparents being her father, her mother, and Mary Ranson, probably her mother's sister. Their second child, Joseph Henry, was born on 25 May 1847 at 12 Camden Cottages; his godparents were Henry Cole, Richard Cundall, the publisher's elder brother, and Elizabeth Goodchild Cundall, his sister.

Cole's career on its upward path now took him away from Cundall; but many years later, in 1881, Cundall dedicated his book *On Bookbindings Ancient and Modern* to Sir Henry Cole, 'my earliest and kindest instructor on all questions of art, with sentiments of deep gratitude'.

One small contemporary pen-picture brings Cole and Cundall briefly to life. In a Chiswick Press Ledger now in the British Museum, a note by one of Charles Whittingham's daughters runs as follows:

'One week-day about 1842 Henry Cole came down to Chiswick with a young man, and rather blusteringly inquired for Mr Whittingham; Mr Cole was rather stout and seemed annoyed because I happened to notice him. So I pointed back and told him that Mr Whittingham was in there. Then he went to him, and this young man, Mr Joseph Cundall, who was a red-haired young chap, was introduced, sometime after which he came down one Sunday and brought his wife with him, who, to my Mother's annoyance, wished for a glass of sherry wine. It seems that they expected it would be on the table, but was disappointed because it was not there, and another thing it was on a Sunday when the Father was at home, so some was got and they had it. Some time after this business began between them, and Felix Summerly's Home Treasury was commenced with an edition in modern type of Nursery Songs.'[2]

In 1845, Cundall published the first of his illuminated gift books, *A Booke of Christmas Carols*. It is a charming small book of 32 pages with borders drawn on stone by John Brandard, an artist who specialized in music covers, colour-printed lithographically by M. & N. Hanhart.

The text was overprinted, in Old Face, by the Chiswick Press. The book appeared in at least four different decorative styles of binding, probably all manufactured by Remnant & Edmonds: one was a highly elaborate white paper binding in the French romantic style, with gold embossing and colour printing, comparable to that on Noel Humphreys' *The Illuminated Calendar*, also published in 1845. It also appeared in at least two different colours of flock paper with a gold thread; and in both red cloth and crimson leather, blocked in gold and blind. One of the original drawings for this book is reproduced on p 11 in colour.

Sometime in the 1840's, Cundall was appointed as publisher to the Etching Club[3]: the first book they published was *The Deserted Village* in 1841, but the only ones to carry Cundall's name as publisher seem to have been Gray's *Elegy*, 1847, and *L'Allegro* in 1849. Both works were dedicated to the Queen, and five of the eight

---

[1] Details of Cundall's marriage and children's births, etc., are taken from his family Bible still in the possession of the family.

[2] BM Add. MSS 43986 f.131r.

[3] For a brief description of the Etching Club, see Basil Gray, *The English Print*, 1937, p 105.

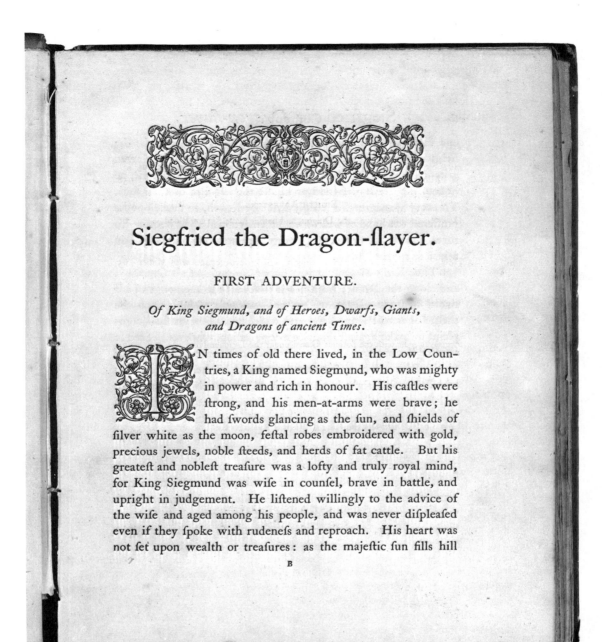

Siegfried the Dragon-flayer.

FIRST ADVENTURE.

Of King Siegmund, and of Heroes, Dwarfs, Giants, and Dragons of ancient Times.

IN times of old there lived, in the Low Countries, a King named Siegmund, who was mighty in power and rich in honour. His caftles were ftrong, and his men-at-arms were brave; he had fwords glancing as the fun, and fhields of filver white as the moon, feftal robes embroidered with gold, precious jewels, noble fteeds, and herds of fat cattle. But his greateft and nobleft treafure was a lofty and truly royal mind, for King Siegmund was wife in counfel, brave in battle, and upright in judgement. He liftened willingly to the advice of the wife and aged among his people, and was never difpleafed even if they fpoke with rudenefs and reproach. His heart was not fet upon wealth or treafures: as the majeftic fun fills hill

B

Siegfried the Dragon-Slayer, 1848. The first text page. Reduced.

contributors (Cope, Horsley, Townsend, Tayler and Redgrave) were Home Treasury illustrators. Whether Cundall took any financial risk, or any part in the production of the books, is not known: it is probable that he merely sold them on a commission basis.

Between 1845 and the beginning of 1849, Cundall published two more small illuminated books, *The Creed, the Lord's Prayer and the Ten Commandments*, and *Words of Truth and Wisdom*, both chromolithographed by F. Dangerfield, and between 40 and 50 books for children, including two of the earliest English editions of Hans Andersen: *A Danish Story-Book*, 1846, and *The Moon's Histories*, 1848.

In 1847 Cundall also launched a monthly magazine for children, *The Playmate*, a curiously dull-looking publication (except for its beautiful colour-printed cover, see p 24) which survived for only a year. In 1846, the Home Treasury was acquired by Chapman & Hall (their list, dated 1 January 1847, announced that the series was now their property) who reissued some titles with their own imprint. Some other Cundall books appeared with an imprint shared with Chapman & Hall.

The energetic firm of Chapman & Hall were already producing children's books in direct competition with the Home Treasury, of which some are illustrated in the present writer's *Victorian Book Design*, 2nd edn, pp 50, 55 and 214. They are in the 'ordinary' style of the 1840's; Cundall's are in the purer and simpler Chiswick Press style. Despite all the idealism of Henry Cole and Cundall, it was the Home Treasury that went to the wall. Cundall had to learn one of the harder truths about publishing, that it is not enough to produce handsome and worthwhile books; they must also be sold. When he got into financial difficulties in the late 1840's, it must be assumed that it was simply because his sales were not keeping pace with his production.

It is in 1849 that we find the first book produced by Cundall, credited to him but with another publisher's imprint.[1] This was a remarkable small book, *Songs, Madrigals and Sonnets*, published by Longman, Brown, Green & Co. in 1849. It consists of 72 pages, in a page size of only 137 × 102 mm. The pages are of stiff white card, and 64 of them are printed in colours from wood. The preface, signed 'J. C., Camden Cottages, December 1848' is as follows:

'IN selecting the *Songs*, *Madrigals* and *Sonnets* in this little volume, I have merely endeavoured to choose those which are the most familiar and the most pleasant. The only arrangement that has been attempted is like that of a garden; flowers of all ages and all hues are set side by side, each adding to the general beauty.

'As *Madrigals* and *Sonnets* are of Italian origin, borders of an Italian character of design have been thought the most appropriate decoration.'

The poems are set in the equivalent of 8 pt Caslon Old Face italic, by the Chiswick Press. When it was published, the commercial use of colour printing to decorate books was still a novelty. Chromolithography, pioneered in England for book decoration by Owen Jones, had been in commercial use for about ten years, and wood-block printing in colours commercially for roughly the same period;

---

[1] In 1845, Cundall published *Sixty Etchings of Reynard the Fox* and Longman's published *Reynard the Fox*, translated by S. Naylor. Both volumes were decoratively designed, with uniform exotic bindings in white cloth. It may be that this is the true forerunner of Cundall's freelance work for other publishers, although no proof of this has been found in surviving Longman ledgers.

The Creed, etc, 1848. Cover in blue and gold flock paper on boards.

Mine be a cot beside the hill;
A bee-hive's hum shall soothe my ear;
A willowy brook that turns a mill,
With many a fall, shall linger near.

A Book of Favourite Modern Ballads, n.d. (c. 1865). Reduced. In original, both pages are in colour.
The ornamental designs on this title-page opening are by Albert H. Warren, the drawings
by Birket Foster, the engraving on wood and printing in colour by Edmund Evans.

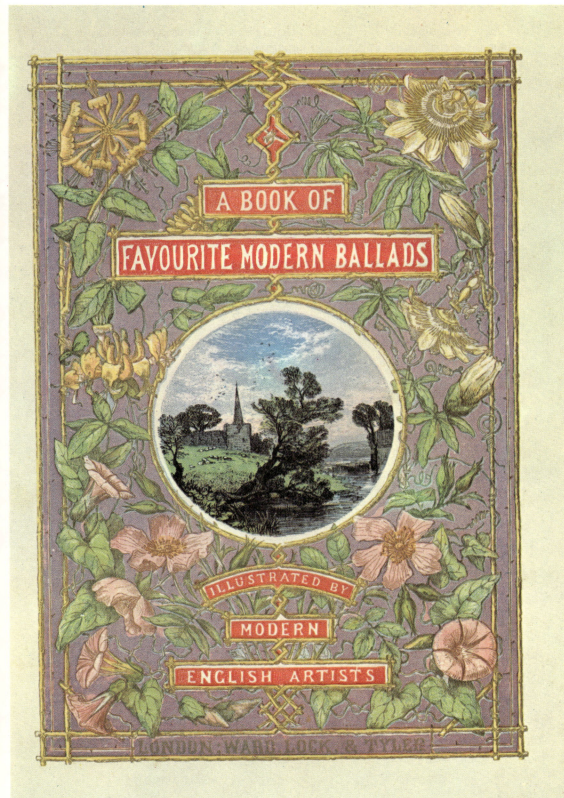

A BOOK OF
FAVOURITE MODERN BALLADS

ILLUSTRATED BY

MODERN

ENGLISH ARTISTS

LONDON: WARD, LOCK, & TYLER.

but they were still rarities. Perhaps the first English book to use wood-blocks in colour for decoration (as opposed to illustration) was *A Bridal Gift*, printed and published by David Marples of Liverpool in 1847 – but not more than two colours were used in any designs and the artistic standard was not high. Cundall's *Songs, Madrigals and Sonnets* – which was not only edited but conceived and therefore 'designed' by him – is a work of art in its small way, and genuinely original. It is not modelled on any particular period of printed book design: the closest parallels are in French books of the sixteenth century. From Messrs Longman's accounts for the book, which survive, we know that John Absolon was paid £36 'for drawings and designs', but we do not know what his sources were; they seem Pompeian. Cundall might well have gone to Henry Shaw, from whose historical researches he often took inspiration, e.g. the Holbein prototype for the Home Treasury covers, but the closest I can find is a page of 'Pannells in the Marble from the facade of the Certosa di Pavia, beginning of 16th Century', in the *Encyclopaedia of Ornament*, 1842.

Songs, Madrigals and Sonnets, like many other notable books, was not a commercial success. It is melancholy to read in Messrs Longman's accounts that only 250 copies had been sold by June 1850. In January 1851 40 bound copies and 1,642 copies in sheets were disposed of to H. G. Bohn at about 1s 8d each. One-quarter share in the profits was to be paid to Joseph Cundall, but there were none.

The Cundall's third child, Herbert Minton (christened after the potter, who had manufactured the 'Tesselated Pastime' and the Terra-Cotta Bricks for the Home Treasury[1]) was born at 12 Camden Cottages on 25 August 1848. His godparents were George Bell, the publisher, Lawrence Drake Cundall, the youngest son of Joseph's father's second marriage, and Eliza Randell.

Probably in 1849, Joseph Cundall suffered bankruptcy, and moved both his home and office. Two details only of the bankruptcy have so far been found, both in the Chiswick Press account books, now in the British Museum: in a ledger of 1849, both Cundall and H. M. Addey (who around that date became Cundall's partner) are entered as bankrupt, and after the publisher William Pickering's death in 1854, Cundall appears in the Bad Debts Column as a debtor for £6 13s 11d. But whatever the details of his bankruptcy, it does not seem to have interrupted Cundall's flow of work.

Cundall's new home was at 3 Bellina Villas, in Kentish Town, where he remained until 1857 or 1858. Here his fourth child, Edmund, was born on 27 January 1850; his godparents were John Ranson (presumably his uncle, see the family tree on pp. 2-3) and Henry and Emma Hoar. Edmund joined the Navy and died at sea as a midshipman.

[1] Cundall asked Minton to be godfather: in a letter preserved in the family, dated 18 Sept 1848, Minton replied '. . . I have for some years declined standing as Godfather to any but *very near relations*, & have offended several friends by my refusal; I hope however you will have more sense than to be offended by my refusal—if you call him by my name, I will gladly give him a cup and in addition, when I see the lad, I will give him a thump, and assure him that there never was a Herbert who prospered in this world, without having been well knocked about.

> I remain,
> My dear Sir,
> Yrs very truly
> H. Minton'

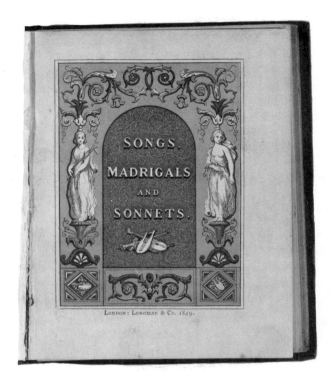

Songs, Madrigals and Sonnets, 1849. Title-page printed in black, yellow, red, emerald green, blue and brown from wood by Gregory, Collins & Reynolds and/or the Chiswick Press, and a text spread. Reduced.

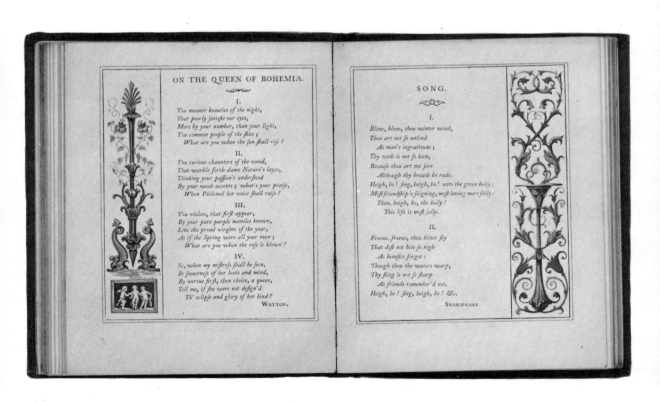

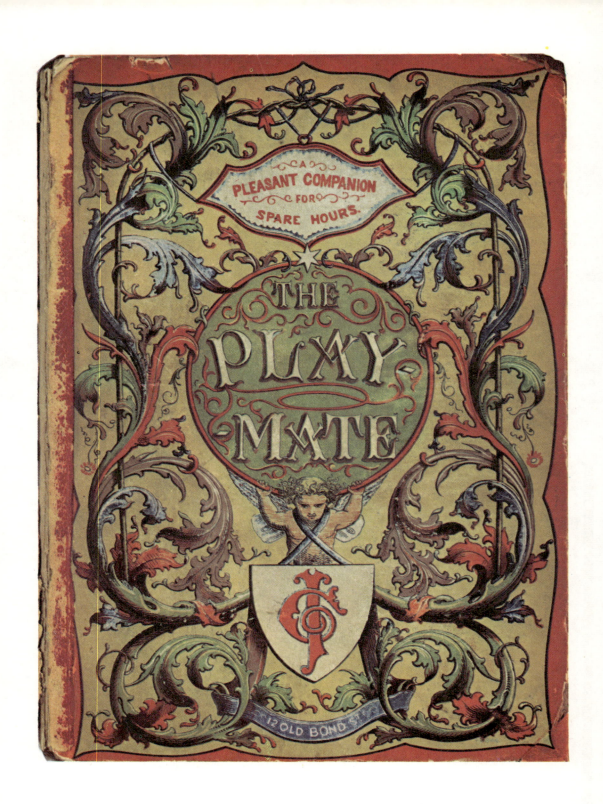

A
PLEASANT COMPANION
FOR
SPARE HOURS.

THE
PLAY-
MATE

12 OLD BOND ST

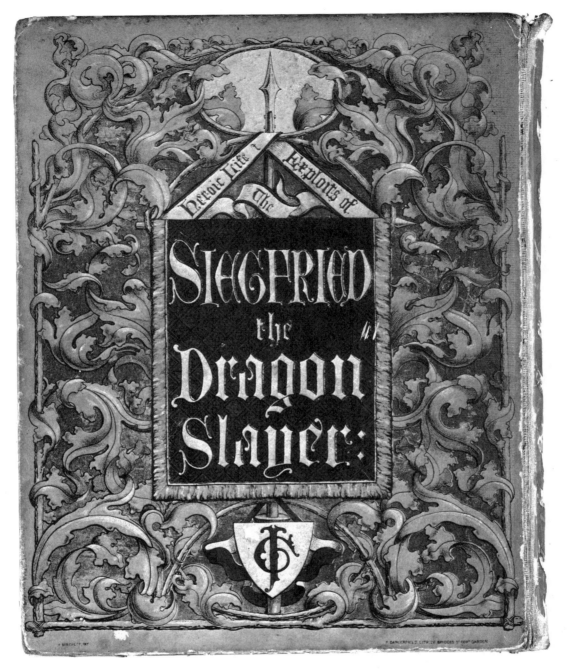

Siegfried the Dragon-Slayer, 1848. Cover designed by R. Burchett and printed on paper boards in colours from stone by F. Dangerfield. Reduced.

(*facing*) *The Playmate*, 1847. Cover designed by R. Burchett and printed on paper boards in black, red, blue, purple, green and yellow from wood by Collins & Reynolds. Note Joseph Cundall's monogram on shield, on both covers.

II. 21 OLD BOND STREET 1849-52

OME time in 1849, Cundall moved his office to 21 Old Bond Street, and entered into partnership with Addey. Addey himself remains a completely mysterious figure about whom little seems to be known beyond the names of books he announced for publication in various advertisements.

I have so far found no books bearing the imprint 'Cundall & Addey' dated before 1851, but in that year I have listed 14 titles (10 for children). An advertisement in the *Publishers' Circular*, vol. XIV, no. 341 dated '21 Old Bond Street, December 1, 1851' is headed 'Messrs. Addey & Co. (late Cundall & Addey)'. I have noted two books only, with the imprint 'Addey & Co. (late Cundall & Addey)', dated 1852.

At the end of his partnership with Addey, apparently sometime in 1852, Cundall moved to 168 New Bond Street; Addey remained at 21 Old Bond Street at least until 1854. By 1856 Addey was operating from Henrietta Street, Covent Garden; but I have seen only two 'Addey & Co.' books dated after 1856: *The Ocean Child, or Showers and Sunshine*, by Mrs Harriet Myrtle, and Harrison Weir's *The Book of Animals*, both dated 1857. (*The Whist-Player*, by Lt. Col. ★★★★, Addey & Co., 1856, was republished in its second edition in 1858 by Chapman & Hall).

From the time of his bankruptcy, Cundall worked more and more as an editor and book-producer for other publishers. Sometimes he would have been paid a flat fee for his work on a book, but no doubt there were other arrangements whereby he would contract for a share of the profits, as with *Songs, Madrigals and Sonnets*. He also continued to publish a certain number of books himself; and not a few books have the imprint 'published for Joseph Cundall by So-and-So' which means probably that the other publisher handled the distribution and selling side, on commission; the risk would be taken by Cundall. Where the imprint names another publisher jointly with Cundall, one assumes that the risk was shared. During 1849 and 1850, Cundall published several titles in association with Bogue, at least one with Grant and Griffith, and one was published for Cundall by Sampson Low. In the same period, of two years, I have noted eight books published with Cundall's imprint alone.

Of the books published from 21 Old Bond Street, the three most distinguished were *The Peacock at Home*, Joseph Cundall, 1849, *A Chaplet of Pearls*, Cundall & Addey, 1851, and *Choice Examples of Art Workmanship*, Cundall & Addey, 1851.

The Peacock at Home, which had been first published by J. Harris in 1807, and later by N. Hailes, was an illuminated book for children, with fine typography by the Chiswick Press. *A Chaplet of Pearls* was an illuminated anthology for adults, with borders printed by chromolithography round text printed letterpress, in black letter, by the Chiswick Press. The decorations, drawn by Mrs Charles Randolph, are good but unremarkable: what is intriguing is the choice of poems, in French and Spanish as well as English, which argue an editor with a far more

Choice Examples of Art Workmanship, 1851. Cover printed in red, yellow and other colours from wood on paper boards. Note circular brass knobs at corners, to hold design off surfaces and prevent it getting rubbed. Reduced.

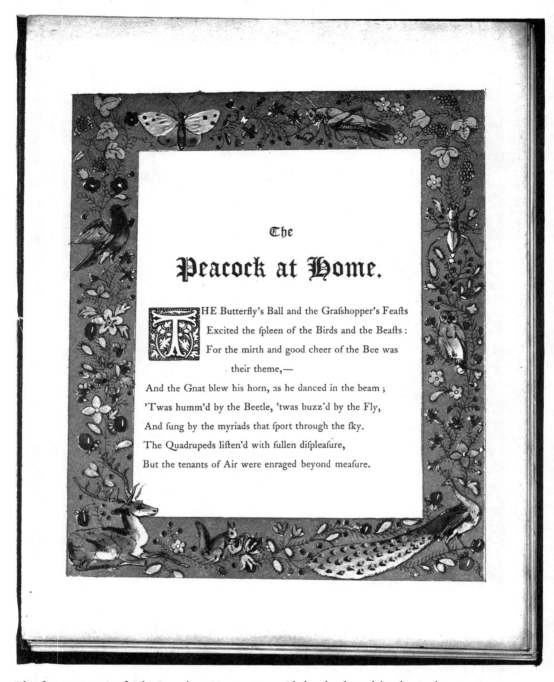

The
Peacock at Home.

THE Butterfly's Ball and the Grafshopper's Feafts
Excited the fpleen of the Birds and the Beafts :
For the mirth and good cheer of the Bee was
　　. their theme,—
And the Gnat blew his horn, as he danced in the beam ;
'Twas humm'd by the Beetle, 'twas buzz'd by the Fly,
And fung by the myriads that fport through the fky.
The Quadrupeds liften'd with fullen difpleafure,
But the tenants of Air were enraged beyond meafure.

The first text page of *The Peacock at Home*, 1849, with hand-coloured border, and text printed by the Chiswick Press. Reduced.

A Rhyme for Goode Friday.

Sith þ worldes Kinge, for man's iniquitie
Dyed on þ crosse man reared on Calvarye,—
Thatte hard and felon dethe of infamie
Ever in Earthe, and Heav'n's eternity
Dwelles, wyth a deare and bitter memory.
Makes Life one dethe, and Dethe an ecstasye,
Makes paine a crowne, and ease uneasy crosse;—
Makes sorowe ioye, and ioye all vanitye,
Makes losse a gaine, and gain severest losse.
Makes poortythe wealthe, and golde lesse
            worth than drosse,
Makes peace a strife, and grame an amnestye,
Makes toile a reste, and reste an industrie,
Makes Conqueste foile, and foile a victory.

A Chaplet of Pearls, 1851. A secular 'illuminated' gift book, with borders in chromolithography and text probably printed letterpress by the Chiswick Press. Reduced.

virile and 'modern' taste than was usual with anthologists of the nineteenth century. *Choice Examples of Art Workmanship* was a book of engravings on wood 'from objects of Ancient and Medieval Art exhibited at the Society of Arts, selected and drawn under the superintendance of Philip De la Motte'. Like the works of Henry Shaw, it aimed to show good designs of the past for comparison with those designs of the present (e.g. those exhibited in 1851 at the Great Exhibition), which were thought deplorable by Henry Cole, Owen Jones and others whom Cundall admired. As a book, its most remarkable feature was a superb decorative binding, printed in four colours from wood on paper, and embellished with brass knobs on front and back to hold it off the table or shelf if lying flat. These covers were included inside the book as end-papers for the special limited edition of a few copies 'printed on vellum, carefully illuminated and finished by Mr De la Motte, bound in velvet, price twelve guineas'.

It is not known that Cundall ever commissioned illustrations from 'Dicky' Doyle, or, indeed, ever met him, but in 1851 Cundall and Addey published *The Story of Jack and the Giants*, with some of Doyle's finest illustrations. This book was actually commissioned by the wood-engraving firm of the Dalziels, and it is a sign of Cundall's prestige at that time as a publisher of children's books that he was entrusted with this publication by the experienced brothers Dalziel.

The books so far mentioned do not seem to have been reprinted quickly and were probably not, therefore, money-makers; but in 1850 Cundall published, in association with Grant and Griffith, *A Treasury of Pleasure Books for Young Children*, which judging by the number of its reprints, in varying forms, must surely have been really profitable. It must have been planned by Cundall, since the foreword 'To my dear children, Maja, Harry and Herbert' was signed by him. A special feature was the decorative cover design and end-papers by Owen Jones, which Cundall recommends in his preface. Handsome as they are, one must wonder what appeal they had to children in 1850; but the rest of the book was less pretentious, with more than 100 engravings on wood, in an open and simple style, after drawings by John Absolon and Harrison Weir.

In this book, as Iona and Peter Opie point out in *The Classic Fairy Tales*, 1974, Cundall made a change in the *Story of the Three Bears*, which has stuck: he turned the old woman into a little girl. She did not, however, get her name 'Goldilocks' until about 1904.

There were many reprints of the book, and the individual stories were also issued separately in paper covers, at 6d plain or 1s coloured, and remained in print for many years.

Harry's Ladder to Learning, which Cundall published in six parts between 1849 and 1850, and then in volume form with David Bogue, was a similarly simple book for young children, which looks as if it also must have been commercially successful.

The Cundall's fifth child, Rose, was born at Bellina Villas on 15 November 1851; her godparents were Thomas Edlin, Anne Hawkes and Ellen Ranson.

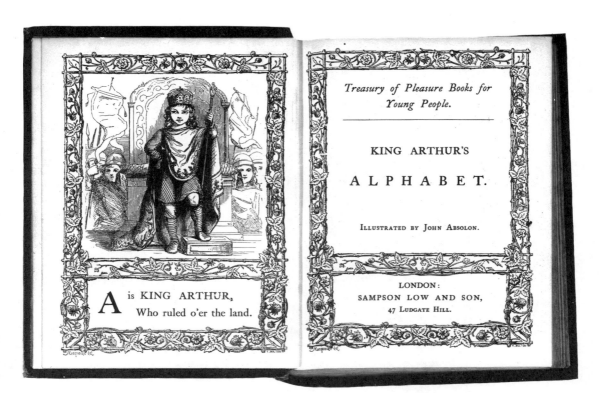

KING ARTHUR'S

A L P H A B E T.

ILLUSTRATED BY JOHN ABSOLON.

LONDON:
SAMPSON LOW AND SON,
47 LUDGATE HILL.

A is KING ARTHUR,
Who ruled o'er the land.

The Blackbird and the Thrush,
And charming Nightingale,
Whose sweet jug sweetly echoes
Through every grove and dale;

The Sparrow and Tom-Tit,
And many more, were there;
All came to see the wedding
Of Jenny Wren the fair.

The Bulfinch walk'd by Robin,
And thus to him did say,
" Pray mark, friend Robin Redbreast,
That Goldfinch, dress'd so gay;

What though her gay apparel
Becomes her very well,
Yet Jenny's modest dress and look
Must bear away the bell."

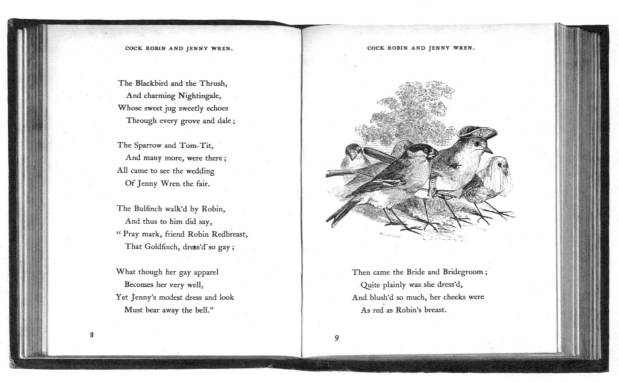

Then came the Bride and Bridegroom;
Quite plainly was she dress'd,
And blush'd so much, her cheeks were
As red as Robin's breast.

8

9

A Treasury of Pleasure Books, 1850. Two openings. Reduced.

CUNDALL, DOWNES & Cº   PHOTOGRAPHERS

Joseph Cundall's first wife, Sarah. Enlarged from a Cundall, Downes & Co.
carte-de-visite photograph.

III. 168 NEW BOND STREET 1852-68

N his new premises at 168 New Bond Street, to which he moved in 1852, Cundall continued to work as editor, author, publisher and book-producer, and also entered the photographic business. He had already in 1847 become one of the twelve original members of the Photographic Club,[1] and in 1853 became a founder-member of the Photographic Society, now the Royal Photographic Society of Great Britain. At his new premises, named 'The Photographic Institution', the second photographic exhibition ever held in Great Britain was arranged by Philip H. De la Motte, whose manual, *The Practice of Photography*, was published by Cundall in 1853. Cundall himself wrote and published in 1854 a small 32-page booklet *The Photographic Primer*, bound in yellow paper with an elaborate design printed in red from a woodcut by John Leighton, and containing a remarkable frontispiece described as 'A Facsimile of a Photographic Picture of Birds, showing the Difference of Tone produced by Various Colours': it is in fact a Leighton Brothers woodcut, printed in several tones of brown, and embossed.[2]

Various other manuals of photography, and collections of photographs, were published from the Photographic Institution. One of the most important was Philip De la Motte's classic *The Progress of the Crystal Palace, Sydenham*, in 1855, which contained far better photographs of the inside of the Crystal Palace than any that had been previously taken. In *Twenty Views in Gloucestershire*, 1854, the photographs were taken by Cundall himself, according to a statement in the book, 'at the request of an eminent engineer [Brunel?], to be laid, in Evidence, before a Committee of the House of Commons. They were produced by the Collodion process, in the third week of March, and were all developed in a Post-chaise.'

One of Cundall's clients at 168 New Bond Street was Lewis Carroll, who when in London regularly entrusted Cundall with the printing of his negatives, which were also stored by Cundall.[3]

In the 1850's, Cundall's photographic business was trading as Cundall, Howlett & Co. (Robert Howlett took the famous photograph of Brunel chewing his cigar in front of the *Great Eastern*'s launching chains, and the photographs used by Frith when painting *Derby Day*), then as Cundall, Howlett & Downes, then Cundall, Downes & Co., and, by 1866, or earlier, Cundall & Fleming. Details of Cundall's photographic businesses are lacking. The back of a Cundall, Downes & Co. *carte de visite* photograph of Cundall's wife (see p 32) shows that they were appointed as Photographers to the Queen, and had studios in Kensington, Bury St Edmunds, Colchester and Bedford. Since Cundall was so deeply involved in photography, it is strange that very few photographs of Cundall himself have come

[1] The twelve names are given in Gernsheim's *History of Photography*, 1969.

[2] There are copies in the British Museum and the St Bride Printing Libraries.

[3] H. Gernsheim, *Lewis Carroll, Photographer*, 1949, p 27.

to light: we know of only two taken in middle age (see pp viii, 46) and one as an older man (see p 88) and we do not know who took the first two.

In 1856, on the instructions of the Queen, Cundall and Howlett took photographs at Woolwich of soldier heroes newly returned from the Crimea.[1] One of these, specifically credited to Cundall in *The Photographic Album*, 1857, is shown on p 35.

Between 1866 and 1870, Bell & Daldy published a series of handsome art books, all illustrated with pasted-down photographs by Cundall & Fleming, and all printed at the Chiswick Press. The series may well have been planned and edited by Cundall. Various other books containing photographs by Cundall & Fleming, but not necessarily designed by Cundall, are described in the check-list. It was not until the 1880's that the reproduction of photographs by normal printing processes (lithography or letterpress) became feasible, and superseded the pasting-down of photographs onto book pages.

We can deduce Cundall's activities in book publishing at 168 New Bond Street only from the evidence of the books themselves. He continued to do a certain amount of publishing under his own name, but more and more he seems to have been producing books for other publishers.

A monogram of his initials was designed by W. Harry Rogers, and frequently appeared on the verso of Cundall's title-pages (see p 39). Other monogram designs by Rogers for Cundall are shown on pp 71 and 92.

Most of the books Cundall produced were illustrated, either in colour or black and white, usually by wood-engravings. One assumes that he proposed the idea, commissioned all the contributors, took care of the entire production (which would mean that he virtually designed the book, or certainly acted as art director) and took a percentage of the profits. Evidence of Cundall as the printer's customer is given on many pages of the Chiswick Press ledgers, and by statements in Edmund Evans's *Reminiscences*, e.g. 'For Joseph Cundall I engraved and printed the illustrated pages to *Sabbath Bells* . . .' and 'I engraved and printed for Joseph Cundall *Favourite Modern Ballads*'. The publishers he worked for or with included Thomas Bosworth; George Bell (later, Bell & Daldy); Routledge; W. Kent; Ward, Lock & Tyler; Bickers; and Sampson Low, for whom for many years he had a close and regular connection. Mr Marston, Sampson Low's partner, wrote of Cundall in his reminiscences *After Work*, 1904: 'He assisted us in bringing out many of the finely illustrated works for which our house has been well known. I am glad to recall his name and cheerful presence. . . .'

The six colour-plate gift books known to have been produced by Cundall, apart from his 'illuminated books', are all in their own ways good examples of book design and production, comparing favourably with other similar works produced in England during the same period. They were unpretentious, popular, intended for reading as well as looking at, and not, like so many books of the period, evoking any style of the past. Each was designed as a whole, was contemporary in feeling, basically simple in treatment, and not overloaded with meaningless decoration. In three of them, the letterpress text was set and printed by the Chiswick Press: in two others, by Edmund Evans, and in one, by G. Barclay.

In 1853, Cundall produced *The Poets of the Woods* for T. Bosworth, and *Poetry of the Year* for George Bell. *The Poets of the Woods* was joined in 1854 by a companion

[1] *Illustrated London News*, 12 April 1856, pp 369–370. I am indebted to Miss Allum of the National Army Museum for this reference.

Highlanders, photographed in 1856 by Joseph Cundall.
From *The Photographic Album*, 1857. *Courtesy The Royal Photographic Society.*

35

volume, *Feathered Favourites*. Both appeared in elegant cloth bindings, blocked in gold, and contained twelve circular chromolithographs by M. & N. Hanhart, after paintings by the animal painter Joseph Wolf (1820-99). Wolf's biographer, A. H. Palmer, states that the original drawings for these plates 'are gems, glowing in the most delicately harmonious colouring' and continues: 'As for the chromolithographs, they are pretty, it is true; a few are very pretty; but they do not fairly represent Joseph Wolf; and the marvellous rococo borders that surround them would offend the taste of a fairly well-educated cheesemonger'.[1] Palmer's views can be tested against modern taste, whether of cheesemongers or not, by referring to the reproduction on p 37. He goes on to relate that the original drawings, after disappearing in the failure of an auctioneer, subsequently turned up in a city sale-room and were bought for a nominal sum by H. E. Dresser, the ornithologist; it would be interesting to know if they still exist.

Neither *The Poets of the Woods* nor *Feathered Favourites* contains any mention of Joseph Cundall, but they are identified as his in *Sabbath Bells Chimed by the Poets*.

Poetry of the Year, published by Bell in 1853, was illustrated with 22 plates printed in chromolithography by Hanhart, pasted down on heavy cartridge; the text was printed by G. Barclay. This was one of the first books, if not the first in England, in which landscape paintings were reproduced by chromolithography, albeit in a much reduced size and for decorative rather than instructional reasons. There was a second edition, with some alterations, published *c.* 1867 by Charles Griffin.

Then, in 1856, appeared *Sabbath Bells Chimed by the Poets*, probably one of the most successful books (judging by the numerous editions which appeared) that Cundall ever produced. The first edition was published by Bell & Daldy, although it is probable – from Edmund Evans's statement quoted on p 34 – that Cundall was the main proprietor. It was the first book with illustrations printed in full colour from wood-engravings by Edmund Evans: the drawings were by Birket Foster, who until 1846 had been a fellow-apprentice with Evans at Landell's. The typography and letterpress printing of the text, specially mentioned in Cundall's editorial note, were by the Chiswick Press: and the Chiswick Press initial letters, used for the beginning of each new poem, were in some copies coloured by hand, indicating perhaps that this luxury was still not more expensive than colour printing. Cundall's monogram, hand-coloured, appeared on the verso of the title-page; it was possibly commissioned for this book, but seems to have first appeared in *The Pleasures of Hope*, one of Sampson Low's 'Present Books', in 1855. *Sabbath Bells* is a very pretty book. The second edition appeared in 1861, also published by Bell & Daldy, but a 'new edition' (possibly the same sheets as the second edition with a cancel title) appeared in the same year with Cundall's name on the title-page. The book went on appearing, from different publishers, with the same blocks but not always printed by Evans, for many years. A line at the foot of the contents page 'Selected by the Editor of "The Poetry of the Year", "The Poets of the Woods", etc.,' coupled with Cundall's monogram, identifies the origin of all these books.

Cundall's next gift book (discounting R. A. Willmott's beautiful edition of *The Poems of Oliver Goldsmith*, 1859, see p 84) was *A Book of Favourite Modern Ballads*, published by W. Kent (late D. Bogue) in 1860, containing illustrations by numerous artists, engraved on wood by Edmund Evans, and borders and decorations by

[1] A. H. Palmer, *The Life of J. Wolf*, 1895.

The Goldfinch, a colour plate from *The Poets of the Woods*, 1853, illustrated by Joseph Wolf.

Albert Warren on every page, printed in gold. In the first edition, the illustrations were printed in black, but in about 1865 the book was republished by Ward, Lock & Tyler with all the illustrations printed in six or eight colours, including the magnificent double-spread title (see pp 20-1). The colour-printed version was subsequently issued in two halves, as *Choice Pictures and Choice Poems*, and *The Illustrated Poetical Gift Book*. All versions of this book, both monochrome and coloured, were extremely handsome.

The sixth of Cundall's colour-plate gift books was *Shakespeare's Songs & Sonnets*, published by Sampson Low in 1862. It was much bigger than the others, being a folio, illustrated with ten chromolithographs after paintings by John Gilbert printed in ten or twelve colours by Vincent Brooks, which for brilliance and brightness of colouring surpass any chromolithographs I can remember: it is difficult to believe that the artist's original paintings could have been any better. The text, and smaller illustrations by Gilbert on wood, were printed by Edmund Evans in two colours.

There were not many other colour-plate gift books during this period (apart from illuminated books) which may be compared with Cundall's, and it is perhaps worth while to enumerate them. They are Dalziels' *Odes and Sonnets*, 1859; *The Poems of Oliver Goldsmith*, 1859 and 1860 (mentioned on pp 36 and 84); *Gems from the Poets*, 1860; *The Art Album*, 1861; *Gems of Nature and Art*, c. 1868; and *Gems of English Art*, 1869.

Of these, *The Poems of Oliver Goldsmith* and *Common Wayside Flowers* (with its sensational cloth and paper binding by Albert Warren and Birket Foster's delicate flower paintings printed in colour by Edmund Evans) are both well enough done to deserve association with Cundall, but have probably nothing to do with him. Both were published by Routledge. *Odes & Sonnets*, also from Routledge, is a pretty book, but does not come near the gracefulness of Cundall's usual touch. *The Art Album* (W. Kent) with colour printing by Evans, and *Gems of English Art* (Routledge) with colour printing by Leighton, both have magnificent, anonymous binding designs but are inferior typographically; and *Gems from the Poets* and *Gems of Nature and Art*, published by Groombridge, both bear witness to the idiosyncrasies of A. F. Lydon as illustrator and Benjamin Fawcett as colour printer. Both are charming, but dated in a way that does not, I think, apply to Cundall's books.

Cundall's main work during the 1850's seems to have been gift books illustrated with wood-engravings in black-and-white, which he produced for various publishers but mostly for Sampson Low. For Sampson Low, he planned the well-known series of small, thin books, measuring 196 × 130 mm., whose only generic title seems to have been 'Illustrated Present Books'. These were nearly all printed extremely well by letterpress, but one, Warton's *The Hamlet*, was printed from copper-plates, and *The Babes in the Wood* was printed in colour from wood by Dickes. Most of them carry Cundall's monogram, and some the imprint 'Published for Joseph Cundall by Sampson Low & Son'; but the exact arrangement between the two firms is not known.

The exact amount of Cundall's contribution to any given book bearing his name or monogram is impossible to determine today, but one may, perhaps, assume that the general look and arrangement, and the selection of the illustrators and embellishers, including the design of the case, would be his responsibility. For several of the anthologies, e.g. *Favourite English Poems*, *Early English Poems*, and *The Poets of the*

SONGS OF THE BRAVE.

THE SOLDIER'S DREAM,

AND OTHER

POEMS AND ODES.

BY

CAMPBELL, WOLFE, COLLINS, BYRON, TENNYSON, AND MACKAY.

ILLUSTRATED WITH TWENTY-SIX ENGRAVINGS,
FROM DRAWINGS BY EDWARD DUNCAN, BIRKET FOSTER, GEORGE THOMAS, ETC

LONDON: SAMPSON LOW, SON & CO. 47, LUDGATE HILL.
MDCCCLVI.

Songs of the Brave, 1856, in Sampson Low's 'illustrated Present Books' series. One of a very few books carrying Cundall's monogram on the title-page.

RHYMES AND ROUNDELAYES

IN PRAISE OF A

COUNTRY LIFE.

ADORNED WITH MANY PICTURES.

SECOND EDITION.

LONDON :
DAVID BOGUE, 86, FLEET STREET.
M DCCC LVII.

Rhymes and Roundelayes in Praise of a Country Life, 1857, printed by R. Clay. Reduced.

Favourite English Poems of the two last Centuries, 1859, printed by R. Clay. Reduced. This and the plate facing show the characteristic style (using Old-Face types) of Cundall's popular anthologies illustrated with monochrome wood-engravings, meticulously machined by Clay.

Elizabethan Age, Cundall was also the editor, which must have involved him in a great deal of literary labour.

In 1855, Bell & Daldy published *Examples of Ornament*, edited by Cundall, which consisted of 24 plates (six in chromolithography) of ornament selected chiefly from items in the various London museums and the new Crystal Palace. It was a small folio, well produced, and came out one year ahead of Owen Jones's much bigger and more important *Grammar of Ornament*, an epoch-making work of reference which has been repeatedly reprinted right up to the present day. It is a sign both of Cundall's interests and his awareness of what colour printing, then hardly twenty years old, could do for art education and scholarship.

During the fifties and sixties Cundall continued to produce children's books, sometimes for Sampson Low, or in partnership with them, sometimes with other publishers, such as Thomas Bosworth, David Bogue, and Bell & Daldy, and sometimes for himself. One of the prettiest was *Songs for the Little Ones at Home*, with 16 colour plates probably printed by Leighton Brothers, originally published by Sampson Low in 1860, and republished by himself with enlarged text in 1863. The plates, after drawings by Absolon and Birket Foster, are unassuming and charming. It may be that Cundall was in charge of all Sampson Low's output of children's books, but this cannot be ascertained. Various books published by them, such as *The Children's Picture-Book of the Sagacity of Animals*, 1862, and *The Wood-Nymph*, 1870, are in his characteristic style, and have been included in the check-list which follows, but the extent of his contribution to them and others cannot be established without further documentary evidence. A heavy gift book of which Cundall was art director for Sampson Low, published in 1867, was *Two Centuries of Song*, edited by Walter Thornbury, Dickens' collaborator on *Household Words* and biographer of Turner; every page was decorated with borders designed by Henry Shaw, printed in red, meticulously machined by Richard Clay. It appeared in various decorated cloth bindings, some of which were encrusted with brass clasps.

The Cundall's sixth child, Edward George, was born at 3 Bellina Villas on 25 September 1856, his godparents being George Ranson, Philip De la Motte, and Mary Drake Cundall. The seventh and last child, Frank, was born at 17 Carlton Hill East, to which the Cundalls had moved probably late in 1857, on 17 January 1858: his godparents were Francis Bedford, his father, and Mrs George Ranson. The Birket Fosters lived nearly opposite, at 12 Carlton Hill East, and the two families became very friendly. In October 1859, shortly after the death of his wife, Birket Foster and the Cundalls went together on a trip through Belgium to the Rhine and back via Paris, accompanied by Foster's wife's niece who later married Edmund Evans. The trip was celebrated by Cundall in doggerel verses, printed by the Chiswick Press. London ratebooks show that Cundall was resident at 17 Carlton Hill East until mid summer 1864[1]: where he moved to then is not known, but in November 1872 he was in Bournemouth (see check list, 1873, *My Lady's Cabinet*). By August 1876 he was at Barneil, Berrylands, Surbiton Hill.[2]

In 1862 Cundall was appointed Superintendent of the *Illustrated Catalogue* of the International Exhibition held that year in London, a job that looks as if it must have

[1] For this information I am indebted to Mr K. C. Harrison, the City of Westminster Librarian.

[2] Note in diary of Cundall's second wife, Emily, still in possession of Cundall family.

THE GREAT WORKS OF

RAPHAEL SANZIO OF URBINO;

A SERIES OF TWENTY PHOTOGRAPHS FROM

THE BEST ENGRAVINGS OF HIS MOST

CELEBRATED PAINTINGS;

WITH THE LIFE WRITTEN BY GIORGIO VASARI, TRANSLATED, WITH NOTES

AND ILLUSTRATIONS, BY MRS. JONATHAN FOSTER.

AND AN APPENDIX CONTAINING A COMPLETE LIST OF THE AUTHENTICATED

WORKS OF RAPHAEL, TRANSLATED FROM PASSAVANT'S

"RAFAEL VON URBINO UND SEIN VATER."

LONDON:

BELL AND DALDY, 186, FLEET STREET.

CUNDALL AND FLEMING, 168, NEW BOND STREET.

1866.

The Great Works of Raphael, 1866. Cundall edited this book and it contained 20 photographs by Cundall and Fleming. The book was printed by Whittingham & Wilkins at the Chiswick Press. Their typography, as shown here, was as much better than Clay's (as shown in the previous two illustrations) as Clay's was better than most other printers at that time. Reduced.

been onerous. The catalogue was published in four volumes, totalling approximately three thousand pages, and was printed by five English printers (Clay, Clowes, Edmund Evans, Petter & Galpin, and Spottiswoode) with additional contributions by Leighton, Vincent Brooks, The Imperial Printing-Office in Vienna, and the Printer to the Court of Prussia. The pages composed in Vienna and Berlin compare most unfavourably, typographically speaking, with the English ones. Cundall's name appears among the Guarantors of the Exhibition for the sum of £100, as do those of Owen Jones and Digby Wyatt. Henry Cole and Richard Clay were down for £300, Day & Son for £1,000 and Miss Burdett Coutts for £3,000.

IV. LAST YEARS 1868-95

UNDALL was appointed in 1866 as 'Agent for Sale of Examples' at the South Kensington Museum (now the Victoria & Albert) and stayed there until 1889. While there, he was from time to time occupied with some of the earliest systematic photography of works of art. The Victoria & Albert Museum holds, for example, 992 half-plate negatives of all the portraits in the National Portrait Exhibition of 1866 and 856 negatives of the 1867 Portrait Exhibition, all made by Cundall's company.[1]

In August 1871, the Lords of the Committee of Council on Education authorized 'Mr Joseph Cundall to proceed to Bayeux to consult with the authorities and endeavour to obtain permission to make a full-sized photographic reproduction of the tapestry. He was successful in his mission, and Mr E. Dossetter, a skilful photographer, was despatched to Bayeux to commence the work, which he completed in the following year.'[2] The photographs, printed in autotype, were published in F. R. Fowke's *The Bayeux Tapestry*, Arundel Society, 1875, a book later republished in Gleeson White's 'Ex Libris' series. The negatives, 12 × 12 in., and a set of prints coloured by hand in 1872 by students of the Art School (now the Royal College of Art) are still in the Victoria & Albert Museum.

The Victoria & Albert Museum's June staff list for 1889 is the last in which both Cundall and his job are mentioned, the job being finally called 'Superintendent for Examples and Publications'.[3] In 1880, Joseph's youngest son Frank, then aged 22, was employed at the Museum in cataloguing the National Art Library, which he did until 1882, and for the next five years he was Assistant Secretary to the various exhibitions held in South Kensington. On 6 February 1891, Frank was appointed as Secretary and Librarian of the Institute of Jamaica, where he remained until his death in 1937. During this time he founded the Art Gallery and Museum in Kingston, Jamaica, was a much respected figure in the literary and artistic life of that country, and was the author of numerous works bearing on its history.

Joseph continued writing and editorial work during his later years: he edited, for

---

[1] I am indebted to Mr John Physick of the Victoria & Albert Museum for this information.

[2] F. R. Fowke, *The Bayeux Tapestry*, 1875, p 10.

[3] I am grateful to Mr Charles H. Gibbs-Smith, Keeper Emeritus of the Victoria & Albert Museum, for this information.

Sampson Low, *Illustrated Biographies of the Great Artists*, in 39 volumes, 1879-91, and was responsible for some of the volumes himself, including that on Hans Holbein; he edited other art books for Sampson Low; he is said, in a note among family papers, to have edited an edition of Don Quixote, but I cannot trace it; and he wrote *On Bookbindings Ancient and Modern*, 1881, which, as previously mentioned, was dedicated to Henry Cole, and which T. J. Cobden-Sanderson mentions in his *Journal* as a book he read before deciding to take up book-binding. Cundall also wrote *Annals of the Life and Work of Shakespeare*, 1886 (Henry Condell, one of Shakespeare's close friends, who with Heminges preserved the Plays and published them after Shakespeare's death, was believed by the family to have been an ancestor) and *Wood-Engraving: A Brief History from its Invention*, 1895.

The Chiswick Press ledgers in the British Museum (BM Add. MS 41936, p 30) show that the Cundall & Fleming partnership was dissolved in 1872. At some later date Cundall moved to Barneil, Berrylands, Surbiton Hill. When he died, on 10 January 1895, he was living at Lyndhurst House, Wallington, in Surrey.[1] His first wife, Sarah Ranson, had died in 1868. In 1870 he married Emily Anne Thompson; she survived him, joined her stepson Frank in Jamaica, and died there in 1911. Joseph's will was proved for £1,142 10s 9d.

The only personal note in his obituaries is that 'despite persistent asthma, he was of a bright and sanguine nature, and this greatly endeared him to his friends'. He does in fact seem to have been a lovable man who made no enemies. Some glimpses of his personal talents can be seen in his first wife's commonplace book, which is still in the possession of his descendants. It contains two or three poems signed J. C., and three pencil drawings signed J. C., one of Sproughton Church, where they were married in 1845, one of another unnamed church, and another entitled 'Death of a Tomtit', portraying two men shooting in a wood, who are named as James Alton and George Ranson, presumably his wife's brother. Several other poems by Cundall exist; he wrote humorous accounts of expeditions and holidays in verse[2] and had some of them printed by the Chiswick Press for private circulation. Both poems and drawings are unremarkable by the standards of the professionals among whom Cundall spent his daily life, as Cundall was probably well aware. His strength was an unswerving devotion to the highest in poetry and art, and an unshakable integrity in all his dealings. He spent his life, not in becoming rich, but in making books and photographs as well as he could, and far better than most. He was no genius; but he was original, and not an imitator. In his thirty-odd years as a publisher he made a contribution to the world of books and art that deserves not to be forgotten.

[1] *The Times* obituary, 21 January 1895, states that he died in Highgate, and contains other inaccuracies.

[2] e.g. the trip with Birket Foster mentioned above, p 42; for quotations, see H. M. Cundall, *Birket Foster*, 1906, pp 77-8.

Joseph Cundall, from a Cundall family album, date and photographer unknown.

TOWARDS A CHECK-LIST OF BOOKS
PUBLISHED, EDITED OR WRITTEN
BY JOSEPH CUNDALL

* marks books that have not been seen: details of these have usually been taken from publishers' lists or advertisements.

† marks Cundall's major or 'special' books

I. 12 OLD BOND STREET 1841-49

1840 Tales of the Kings of England. Stories of Camps and Battle-Fields, Wars and Victories, from the Old Historians, by Stephen Percy (Joseph Cundall) with 8 engravings on wood after John Gilbert. 'To the kind mother of Magnus and Laurence these tales are affectionately dedicated.' This was Mary, second wife of Joseph's father Benjamin. This volume covers from Julius Caesar to the Black Prince.

Charles Tilt, Fleet St. 134× 102 mm. i-viii+232 pp. Printed by Clarke, Printers, Silver St, London.

1841 Tales of the Kings of England (Richard II to Elizabeth I) by Stephen Percy (Joseph Cundall) with 8 wood-engravings after J. Gilbert. Dedicated 'To my Sister Mary'.

Joseph Cundall, 12 Old Bond Street. 136× 105 mm. i-viii+248 pp. Printed by Vizetelly & Co., 135 Fleet St. Price 4s 6d in richly ornamented cloth.

1846. Fourth edn H. G. Bohn and Joseph Cundall (J. Caesar to Elizabeth I). x+468 pp+ 16 wood-engraved plates.

1841 Cousin Natalia's Tales by the translator of 'Little Henry', with 6 illustrations by Miss Fanny Corbaux.

Joseph Cundall, Hailes's Juvenile Library, 12 Old Bond St. 137× 105 mm. iv+158 pp+2 pp advts+6 pl (lithographs). Printed by Clarke, Silver St, Falcon Square. Price 4s 6d.

An advertisement in this book states 'By the same author: *Tales of the Kings of England*'. This implies that Cundall was the author of *Cousin Natalia's Tales* (and the translator of *Little Henry*). Tp. ill. in *Victorian Book Design*, 2nd edn, 1972, p 53.

1841 The Royal Alphabet of Kings & Queens with 24 highly-coloured full-length portraits of the most celebrated Monarchs drawn by J. Gilbert. (Hand-coloured wood-engravings.) Dedicated to Prince Edward.

Joseph Cundall, Old Bond Street.

1843. Second edn. Printed by Clarke, Printers, Silver St, Falcon Square.

1841 Robin Hood and his Merry Forresters by Stephen Percy (Joseph Cundall), with 8 hand-coloured lithographs after John Gilbert. Dedicated to Cundall's sister Eliza.

Tilt & Bogue, Fleet Street. 165× 124 mm. 154 pp. Text printed by Clarke, plates by Day & Haghe. (Advertised in *Cousin Natalia's Tales* as lately published by Hailes's Juvenile Library.) Price 6s 6d; or with plain plates 5s.

1845. Third edn Joseph Cundall.

1850. Fourth edn H. G. Bohn, with same plates as 1st edn. Text printed by J. & H. Cox.

Ill. in *Victorian Book Design*, 2nd edn, 1972, p 52.

1842 The History of Antiquities of Foulsham in Norfolk by Rev. Thomas Quarles, M.A., R.N.

Joseph Cundall, 12 Old Bond Street. Norwich: Charles Muskett. 200× 120 mm. viii+164 pp+advertisement catalogues of James Burns and Tilt & Bogue. No name of letterpress printer. 4 monochrome lithographic plates by Madeley, 3 Wellington St, Strand. 7s 6d.

1842 Cottage Traditions; or, the Peasant's Tale of Ancestry by Jefferys Taylor.

Joseph Cundall. iv+92 pp. 2s. with one illustration.

1843 Edwin Evelyn, A Tale by Miss Jane
Strickland.
The Peasant's Tale by Jefferys Taylor.
Joseph Cundall, Old Bond Street. 165× 100
mm.
The two stories are in one book. First issued
separately at 2s each.
Osborne Collection

1843 The Heroes of England: stories of the
lives of the most celebrated British Soldiers and
Sailors by Lawrence Drake (Lawrence Drake
was Cundall's half-brother).
Joseph Cundall, Old Bond Street. 164× 104
mm. viii+312 pp+8 copperplates. Text
printed by B. Clarke, Silver Street, Falcon
Square. The copperplates, hand-coloured, are
signed 'J. Gilbert'. There are also 8 woodcut
vignette tail-pieces.
c. 1845. Third edn. Printed by B. Clarke. In
cloth 6s 6d.

THE HOME TREASURY
This series, edited by 'Felix Summerly' (Henry
Cole) and published by Joseph Cundall, was
first announced and began appearing in 1843.
It was followed by the Gammer Gurton series
edited by 'Ambrose Merton, Gent.' (W. J.
Thoms). All titles in both series published by
Cundall were, as far as I know, printed by
Charles Whittingham at the Chiswick Press. In
late 1846 both series were acquired by Chapman
& Hall, who reissued several titles, some of
which were not printed by the Chiswick Press.
In some of Cundall's advertisements the books
are numbered, but the books themselves do not
seem ever to have carried numbers. There are
various discrepancies between the descriptions
of books in advertisements and copies seen.
More titles seem to have been published than
were announced; and many titles appeared in
different editions, e.g. first with illustrations
hand-coloured, and then colour-printed, or by
a different artist. A completely accurate bibliographical description of the series may be
impossible, but the following list shows the
earliest dates on copies seen, with the number of
plates these copies contain. Items not dated
have not been seen; details are from contemporary announcements, or copies. Prices are from
the advertisement at the back of *Beauty and the
Beast*, 1843.

Joseph Cundall, 12 Old Bond Street. 165×
120 mm. All printed by Charles Whittingham.

1843 Bible Events. First series. 8 woodcuts
after Holbein. Coloured, price 4s 6d, plain,
2s 6d.

1843 Traditional Nursery Songs. 8 plates.
Coloured, price 4s 6d, plain, 2s 6d.

**1843 Sir Hornbook or Childe Launcelot's
Expedition.** 8 plates by H. Corbould.
Coloured, price 4s 6d, plain, 2s 6d.

**1843 The Pleasant History of Reynard the
Fox.** 40 plates by Everdingen. Price 6s 6d.

1843 Little Red Riding Hood. 4 plates by
T. Webster. Coloured, price 3s 6d, plain, 2s.

1843 Beauty and the Beast. 4 plates by J. C.
Horsley. Coloured, price 3s 6d, plain, 2s 6d.

1843 Tesselated Pastime. A box of variously
coloured Tesserae accompanied by a book of
Patterns, formed out of the Mosaics published
by Mr Blashfield, purposed to cultivate correct
taste in Ornament. 6s. Double Box, 7s 6d.
Mahogany box, 164× 137× 20 mm; diamond-
shaped colour-printed label pasted on sliding lid.
Contains coloured tesserae and booklet:
Tesselated Pastime, with a chapter on some
Uses of the Toy and on Taste in Ornamental
Patterns, by Felix Summerly.
Joseph Cundall, 12 Old Bond Street. 150× 124
mm. Cover+16 pp text+8 plates printed from
wood in black, red, yellow and blue on rectos
only, showing 61 patterns. Printed by C.
Whittingham, Chiswick. (See illustration facing
title-page.)
Courtesy Justin G. Schiller Ltd.

1844 Bible Events. Second Series. 6 2-colour
lithographs after Raffaelle. Coloured, price
4s 6d, plain, 2s 6d.

1844 Bible Events, Third Series. The Life of
Our Lord Jesus Christ. 5 plates after Dürer.

1844 An Alphabet of Quadrupeds. 24 plates
by Berghem, Dürer, Stoop, Teniers, etc. Cloth
bound volume.

1844 Puck's Reports to Oberon. 6 plates by Townsend, Redgrave, Horsley, Cope and Tayler. Incl. The Sisters, Golden Locks, Grumble & Cheery and The Eagle's Verdict. Coloured, price 4s 6d.

1844 The Ballad of Chevy-Chase. 4 plates by F. Tayler. Coloured, price 3s 6d, plain 2s.

1844 The Lively History of Jack and the Beanstalk. 4 hand-coloured plates by C. W. Cope. Coloured, price 3s 6d, plain 2s 6d.

1844 Heroic Tales of Ancient Greece

1844 The Mother's Primer

for descriptions see p 52

1845 Cinderella. 4 plates by J. Absolon. (Plates advertised to be by R. Redgrave.) Coloured, price 3s 6d, plain 2s.

1845 Jack the Giant Killer. 4 plates by H. J. Townsend. Coloured, price 3s 6d, plain 2s.

1845 Rosebud, the Sleeping Beauty. 4 plates by J. Absolon.

1845 Box of Terra-Cotta Bricks. Geometrically proportioned, each brick being one-eighth of the size of the Common brick in its several measurements, and manufactured under Mr Prosser's Patent, by Messrs Minton, of Stoke upon Trent. With Window frames, door and roof. Price 10s 6d.
With booklet: **Architectural Pastime**, with Terra Cotta Bricks, and Building Plans. Edited by Felix Summerly. London: Joseph Cundall, 12 Old Bond Street. 154×116 mm. Cover+ frontispiece+12 pp text+leaf of semi-stiff paper PLANS FOR BUILDING+6 leaves of hand-coloured plans numbered I-V and VII (VI being the frontispiece), bearing imprint 'Madeley, litho, 3 Wellington St, Strand'.

1846 Tales from Spenser's Faery Queen. 4 plates.

1847 The Veritable History of Whittington and his Cat. Chapman & Hall, 186 Strand, and Joseph Cundall, 12 Old Bond Street.

Printed by C. Whittingham, Chiswick. 3 plates by J. Absolon, printed in colours from wood (by Gregory, Collins & Reynolds?). Price in printed paper boards 1s.

Bible Events, Fourth Series. 6 pictures from the Sistine Chapel after Michael Angelo. The four series of *Bible Events* were listed in 1844 as being available in 'One Volume handsomely bound, 10s 6d plain. Splendidly bound, 21s Coloured'.

★A Century of Fables. Selected from Aesop, Pilpay, Gay, La Fontaine, and others with Pictures by the Old Masters.

★The Little Painter's Portfolio. With 10 Coloured and 4 Plain Pictures by Giotto, S. del Piombo, Holbein, Everdingen, and Modern Artists. 7s 6d.

★Colour Box for Little Painters. With 10 best Colours (including Cobalt, Lake, and Indian Yellow), Slab, and Brushes. Hints and Directions, and specimens of Mixed Tints. 6s 6d.

The following list of the Home Treasury appeared in *Fifty years of public work of Sir Henry Cole, K.C.B.*, vol. II, 1884, p 161.

1. *Holbein's Bible Events. First Series.* 8 Pictures. Coloured, 4s 6d. These were coloured by Mr Linnell's sons.

2. *Raffaelle's Bible Events. Second Series.* 6 Pictures from the Loggie. Coloured, 5s 6d. Drawn on stone by Mr Linnell's children and coloured by them.

3. *Albert Dürer's Bible Events. Third Series.* 6 Pictures from Dürer's 'Small Passion'. Coloured by the brothers Linnell.

4. *Traditional Nursery Songs.* 8 Pictures. 2s 6d. Coloured, 4s 6d. Designed: 'The Beggars coming to Town', by C. W. Cope, R.A.; 'By, O my Baby!', by R. Redgrave, R.A.; 'King in the Counting House', by J. C. Horsley, R.A.; 'Mother Hubbard', by T. Webster, R.A.; '1, 2, 3, 4, 5', 'Sleepy Head', 'Up in a Basket', 'Cat asleep by the Fire' (all four by John Linnell).

5. *The Ballad of Sir Hornbook*. Written by Thos. Love Peacock, with 8 Pictures by H. Corbould. Coloured, 4s 6d.

6. *Chevy Chase*. The Two Ballads with Notes and Music. 4 Pictures by Frederick Tayler, President of the Water Colour Society. Coloured, 4s 6d.

7. *Puck's Reports to Oberon*. Four New Faëry Tales. The Sisters. Golden Locks. Grumble and Cheery. Arts and Arms.[1] Written by C. A. Cole. With 6 Pictures by H. J. Townsend. Coloured, 4s 6d.

8. *Little Red Riding Hood*. With 4 Pictures by Thos. Webster. Coloured, 3s 6d.

9. *Beauty and the Beast*. With 4 Pictures by J. C. Horsley, R.A. 2s. Coloured, 3s 6d.

10. *Jack and the Bean Stalk*. With 4 Pictures by C. W. Cope. 2s. Coloured, 3s 6d.

11. *Cinderella*. With 4 Pictures by Wehnert. Coloured, 3s 6d.

12. *Jack the Giant Killer*. With 4 pictures by C. W. Cope. Coloured, 3s 6d.

13. *The Home Treasury Primer*.[2] Printed in Colours. With Drawing, on zinc, by W. Mulready, R.A.

14. *Alphabet of Quadrupeds*. Selected from the Works of Paul Potter, Karl du Jardin, Teniers, Stoop, Rembrandt, etc., and drawn fron Nature.

15. *The pleasant History of Reynard the Fox*. With 40 Etchings by Everdingen. Coloured, 31s 6d.

16. *A Century of Fables*. Selected from Aesop, Pilpay, Gay, La Fontaine, and others. With Pictures by the Old Masters.

17. *The Little Painter's Portfolio*. With 10 Coloured and 4 Plain Pictures by Giotto, S. Del Piombo, Holbein, Everdingen, and Modern Artists. 7s 6d.

[1] This title may never have appeared. It seems to have been replaced by 'The Eagle's Verdict'.

[2] This book was published in 1844 by Longman, as *The Mother's Primer*, by Mrs Felix Summerly (see p 52). It seems never to have appeared in the Home Treasury with Joseph Cundall's imprint.

18. *Colour Box for Little Painters*. With 10 best Colours (including Cobalt, Lake, and Indian Yellow), Slab and Brushes. Hints and Directions and Specimens of Mixed Tints. 6s 6d. Soon after the production of this box the Society of Arts issued a Prize for a colour-box, and obtained one as good as this, which sold for one shilling!

19. *Tesselated Pastime*. A Toy formed out of Minton's Mosaics with Book of Patterns. 6s. Double Box, 7s 6d.

20. *Box of Terra Cotta Bricks*. Geometrically made, one-eighth the size of real Bricks, by Minton, with Plans and Elevations.

This list omits *Heroic Tales of Ancient Greece*, 1844; *Tales from Spenser's Faery Queen*, 1846; and *The Veritable History of Whittington and his Cat*, 1847; these all certainly appeared in the Home Treasury.

It also omits *Michael Angelo's Bible Events* (*Fourth Series*) which was advertised in 1844.

GAMMER GURTON'S STORY BOOKS

Edited by Ambrose Merton, Gent., F.S.A. (W. J. Thoms). Joseph Cundall, 12 Old Bond Street. 162 × 117 mm each.

1843? **1. Gammer Gurton's Garland.** Illustrated by F. Tayler.

2. The Famous History of Sir Guy of Warwick. Illustrated by F. Tayler.

3. A True Tale of Robin Hood. Illustrated by F. Tayler.

4. The Gallant History of Sir Bevis of Hampton. Illustrated by F. Tayler.

5. The Doleful Story of the Babes in the Wood and **The Lady Isabella's Tragedy.** Illustrated by J. Franklin.

6. A Merry Tale of the King and the Cobbler. Illustrated by J. Absolon.

7. The Famous History of Friar Bacon. Illustrated by J. Franklin.

1845 **8. The Blind Beggar's Daughter of Bethnal Green.** Illustrated by J. Absolon.

Swans are graceful birds.   The Cock has fine feathers.

The Goose hisses.   Turkey is good for dinner.

The Duck says Quack!   Men shoot Partridges.

Little Mary's Primer, 1847. *See also* pp8–9 and 66–69.

51

9. **The Romantic Story of the Princess Rosetta.** Illustrated by J. Absolon.

10. **History of Tom Hickathrift the Conqueror.** Illustrated by F. Tayler.

11. **The Mad Pranks of Robin Goodfellow.** Illustrated by J. Franklin.

1845 12. **A Mournful Ditty of the Death of Fair Rosamond to which is added Queen Eleanor's Confession.** Illustrated by J. Absolon.

13. **The Sweet and Pleasant History of Patient Grissell.** Illustrated by J. Franklin.

In gilt paper cover, 6d, or with the picture coloured, 9d. Some, perhaps all, of the plates, were engraved on wood and printed in colours by Gregory, Collins & Reynolds.

The thirteen booklets were published as a bound volume called *The Old Story Books of England*, with 12 illustrations, of which 11 were printed in colours from wood by Gregory, Collins & Reynolds, and one (for *Gammer Gurton's Garland*) was hand-coloured. The illustration for *Sir Guy of Warwick*, although listed, was omitted. Another edition of *The Old Story Books of England* contains 12 stories (*Gammer Gurton's Garland* is omitted). The illustration to *Sir Guy of Warwick* is included, printed in colour. Chiswick Press, dated 1845.

In 1845, the stories of Patient Grissel, the Princess Rosetta and Robin Goodfellow, and Ballads of The Beggar's Daughter, The Babes in the Wood, and Fair Rosamond, were published by Joseph Cundall as *Gammer Gurton's Pleasant Stories & Ballads*, the illustrations not coloured, and printed by Charles Whittingham.

In 1846 or later, the stories of Sir Guy of Warwick, Sir Bevis of Hampton, Tom Hickathrift, Friar Bacon, Robin Hood and The King and the Cobbler, were published by Chapman & Hall as *Gammer Gurton's Famous Histories*, n.d., with the illustrations hand-coloured. This book was not printed by Charles Whittingham.

In 1859, Sampson Low published *The Home Treasury of Old Story Books* (q.v.), in which it is stated 'This Volume includes all the Fairy Tales which were published in the "Home Treasury" edited by Felix Summerly, and all the "Old Story Books of England" edited by Ambrose Merton'. Among the fairy tales were 'Puss in Boots' and 'Peter the Goatherd', which I have never seen mentioned in a contemporary Home Treasury list, nor found in any printed edition of that date.

1844 Heroic Tales of Ancient Greece related by Berthold Niebuhr to his little son Marcus. Translated from the German. Edited by Felix Summerly. Preface dated 12 Oct. 1843. 4 hand-coloured lithographed ills. by H. J. Townsend.

Joseph Cundall, 12 Old Bond St. 168×120 mm. viii+116 pp+4 plates+4 pp advts. Text printed by C. Whittingham, Chiswick. 4s 6d.

This book was published in the 'Home Treasury' series, but was never included in any list or advertisement of the Home Treasury.

1844 The Mother's Primer. A little Child's first Steps in Many Ways, by Mrs Felix Summerly, with a frontispiece by William Mulready.

Longman, Brown, Green and Longmans. 167×125 mm. 32 pp. Printed by Charles Whittingham in red, blue and yellow throughout.

This book is no. 13 ('*The Home Treasury Primer*') in the Home Treasury list in Sir Henry Cole's *Fifty Years of public work, etc*, 1884 (see p 50): Besides being a remarkable example of Chiswick Press typography and colour printing, in the same rich colours as Byrne's *Euclid*, it shows that Henry Cole's wife knew as much about bringing up young children as any twentieth-century mother. She may well have been a guiding force in the whole Home Treasury series. It would be interesting to know why this title was not published by Cundall. It must be included in the present check-list, since it is inconceivable that Cundall did not contribute something to it. *The Mother's Primer* was reproduced in facsimile in 1970 for the Friends of Osborne and Lillian H. Smith Collections in Toronto Public Library, with an editorial postscript by Judith St John.

End of Home Treasury and Gammer Gurton's Story Books.

BOOKS FOR CHILDREN

PUBLISHED BY

JOSEPH CUNDALL, 12, OLD BOND STREET.

Gammer Gurton's Story Books.

NEWLY REVISED AND AMENDED, FOR THE AMUSEMENT AND
DELIGHT OF ALL GOOD LITTLE MASTERS AND MISSES,
BY AMBROSE MERTON, GENT, F. S. A.

1. GAMMER GURTON'S GARLAND.
2. THE FAMOUS HISTORY OF SIR GUY OF WARWICK.
3. A TRUE TALE OF ROBIN HOOD.
4. THE RENOWNED HISTORY OF SIR BEVIS OF HAMPTON.
5. THE DOLEFUL STORY OF THE BABES IN THE WOOD.
6. A MERRY TALE OF THE KING AND THE COBBLER.
7. THE WONDERFUL HISTORY OF FRIAR BACON.
8. THE RARE BALLAD OF THE BEGGAR'S DAUGHTER.
9. THE ROMANTIC STORY OF THE PRINCESS ROSETTA.
10. AN EXCELLENT HISTORY OF TOM HICKATHRIFT.
11. THE MAD PRANKS OF ROBIN GOODFELLOW.
12. A FAMOUS BALLAD OF FAIR ROSAMOND.
13. THE PLEASING HISTORY OF PATIENT GRISSELL.

Each Book is ornamented with flower borders, and illustrated with a picture by an eminent Artist. Price, in gilt paper cover, 6d.; or, with the picture coloured, 9d.

A Cundall advertisement of 1846. Typography by Charles Whittingham.

The Myrtle Story Books.

A STORY BOOK OF COUNTRY SCENES.—*Written for Young Children.*

A STORY BOOK OF THE SEASONS.—*Spring.*

A STORY BOOK OF THE SEASONS.—*Summer.*

A STORY BOOK OF THE SEASONS.—*Autumn.*

A STORY BOOK OF THE SEASONS.—*Winter.*

Each volume is illustrated with four pictures by Absolon, and is handsomely bound in cloth, price 3s. 6d.; or with coloured plates and gilt edges, 4s. 6d.

Story Books for Holiday Hours.

THE TWO DOVES AND OTHER TALES.

THE LITTLE BASKET MAKER AND OTHER TALES.

THE WATER FAIRY AND OTHER TALES.

THE KING OF THE SWANS AND OTHER TALES.

Each Volume is illustrated with four pictures by Absolon; bound in cloth, price 2s. 6d.; or, with coloured plates, 3s. 6d.

New Story Books.

THE GOOD-NATURED BEAR.

WITH 4 PICTURES BY FREDERICK TAYLER, ESQ. 3s. 6d. COLOURED, 4s. 6d.

MEMOIRS OF A LONDON DOLL.

WITH ILLUSTRATIONS.—*Nearly Ready.*

ANECDOTES OF LITTLE PRINCES.

WITH 6 PICTURES BY J. CALLCOTT HORSLEY, ESQ., 4s. 6d. COLOURED, 6s.

Cundall advertisement, 1846, pages 2 and 3.

Felix Summerly's Home Treasury.

OF BOOKS AND PICTURES; PURPOSED TO CULTIVATE
THE AFFECTIONS, FANCY, IMAGINATION,
AND TASTE OF CHILDREN.

1. **Jack the Giant Killer.**
WITH 4 PICTURES BY TOWNSHEND.
2. **Jack and the Beanstalk.**
WITH 4 PICTURES BY COPE
3. **Sleeping Beauty in the Wood.**
WITH 4 PICTURES BY ABSOLON.
4. **Little Red Riding Hood.**
WITH 4 PICTURES BY WEBSTER.
5. **Cinderella.**
WITH 4 PICTURES BY ABSOLON.
6. **Beauty and the Beast.**
WITH 4 PICTURES BY HORSLEY
7. **Chevy Chase.**
WITH 4 PICTURES BY F. TAYLER.
8. **Golden Locks.**
WITH 3 PICTURES BY REDGRAVE, &c.
9. **Grumble and Cheery.**
WITH 3 PICTURES BY COPE, &c.
10. **The Ballad of Sir Hornbook.**
WITH 4 PICTURES BY H. CORBOULD.
11. **The Life of Christ.**
WITH 4 PICTURES FROM ALBERT DURER.
12. **Bible Events, 1st Series.**
WITH 8 PICTURES FROM HOLBEIN.
13. **Bible Events, 2nd Series.**
WITH 6 PICTURES FROM RAFFAELLE.

*Each of these books is handsomely done up in a gilt cover,
price 1s.; or, coloured and bound, 2s. 6d.*

Bound Series.

1. **The Traditional Nursery Songs of England.**
WITH PICTURES BY EMINENT MODERN ARTISTS, 2s. 6d.
2. **Events in Sacred History.**
WITH PICTURES BY HOLBEIN, RAFFAELLE, AND ALBERT DURER, 4s. 6d.

3. **Traditional Faëry Tales.**
   RED RIDING HOOD, BEAUTY AND BEAST, & JACK AND BEANSTALK, 3s. 6d.

4. **Popular Faëry Tales.**
   JACK THE GIANT KILLER, CINDERELLA, & SLEEPING BEAUTY, 3s. 6d.

5. **Ballads and Faëry Tales.**
   CHEVY CHASE, SIR HORNBOOK, THE SISTERS, GOLDEN LOCKS, GRUMBLE
   AND CHEERY, THE EAGLE'S VERDICT, 4s. 6d

6. **The Delectable History of Reynard the Fox.**
   WITH 24 PICTURES BY ALDERT VAN EVERDINGEN, 4s. 6d.

7. **An Alphabet of Quadrupeds.**
   WITH 24 PICTURES BY BERGHEM, DÜRER, STOOP, TENIERS, &c., 4s. 6d.

8. **Tales of the Heroes of Greece.**
   WITH 4 ILLUSTRATIONS, 3s. 6d.

9. **Tales from Spenser's Faerie Queen.**
   WITH 4 ILLUSTRATIONS BY AN EMINENT ARTIST, 3s. 6d.

Standard Story Books.

ROBINSON CRUSOE.
   THE BEST EDITION.—UNABRIDGED.—6 PICTURES FROM STOTHARD'S
   DESIGNS 10s 6d.; Or, Bound in Morocco and Coloured,, 15s.

ROBIN HOOD AND HIS MERRY FORESTERS.
   BY STEPHEN PERCY,—WITH 8 PICTURES BY GILBERT, 3s. 6d.—Coloured, 5s.

JOHN GILPIN.
   WITH 10 LARGE PICTURES BY A YOUNG ARTIST.—In Fancy Cloth, Price 5s.

LITTLE BO-PEEP and other STORIES, in Rhyme.
   WITH 12 PICTURES BY F. L. M.—2s. 6d.—Coloured, 4s. 6d.

The Holiday Library, etc.

THE ORPHAN OF WATERLOO, *by Mrs. Blackford, 5s.*

HOLLY GRANGE, A Tale, *by Madame Emma de K, 5s.*

LEGENDS OF RUBEZAHL, *from Musäus, 5s.*

HEROES OF ENGLAND, *by Lawrence Drake, Third
Edition, Nearly Ready, 5s.*

NOBLE DEEDS OF WOMAN, *by Elizabeth Starling, 5s.*

Cundall advertisement, 1846, last page.

1843 Little Princes by Mrs John Slater. Hand-coloured frontispiece and 5 hand-coloured lithographic plates after J. C. Horsley. Dedicated to Albert Edward Prince of Wales.

Joseph Cundall, 12 Old Bond Street. 165× 120 mm. xii+196 pp. Text printed by C. Whittingham, plates lithographed by Day & Haghe. Price 4s 6d; coloured, 6s.

1844: Joseph Cundall, 12 Old Bond Street; plates in sand & black only, by Day & Haghe.

1850: Henry G. Bohn, York Street, Covent Garden. Title-page in red and black.

1844 †The Passion of Our Lord Jesus Christ portrayed by Albert Dürer. Edited by Henry Cole.

Joseph Cundall, 12 Old Bond Street; William Pickering, 117 Piccadilly; George Bell, 186 Fleet Street; J. H. Parker, Oxford; J. & J. J. Deighton, Cambridge. 194×133 mm. 96 unnumbered pp. Printed by Charles Whittingham (imprint dated 15 July 1844).

This book was printed from stereotypes made from the original woodcut blocks of Dürer, which had been acquired by the British Museum in 1839 (see p 14).

HAZLITT'S HOLIDAY LIBRARY
(Edited by William Hazlitt)

1844 The Orphan of Waterloo by Mrs Blackford. With 4 lithographed pl by John Absolon, printed by Day & Haghe.

Joseph Cundall, 12 Old Bond Street. 175× 108 mm. Printed by Charles Whittingham. 6s 6d bound.

1844 Holly Grange by Madame Emma de K. With 4 lithotint pl.

Joseph Cundall, 12 Old Bond Street. 175× 108 mm. Printed by Charles Whittingham. 6s 6d bound.

1845 Legends of Rubezahl from the German of Musaus. With 4 lithotint pl printed by Day & Haghe.

Joseph Cundall, 12 Old Bond Street. 172× 105 mm. Printed by Reynell & Weight, Little Pulteney Street. 6s 6d bound.

Illustrated on p 59.

1845 Noble Deeds of Woman (2nd edn) by Elizabeth Starling. Price in cloth 5s.

End of Hazlitt's Holiday Library

1845 Ellen Cameron. A Tale (3rd edn) by Emily Rankin.

Joseph Cundall, 12 Old Bond Street. 133× 95 mm. vi+170 pp+hand-coloured frontispiece by J. Erxlehen lith. by Day & Haghe. No name of letterpress printer.

First published in 1829 by Baldwin & Cradock.

THE MYRTLE STORY BOOKS
Each with 4 illustrations by Absolon. Price, bound in cloth, 3s 6d, or with coloured plates and gilt edges, 4s 6d.

Spring (and probably the other three) was issued in cloth printed with one of Cundall's grolieresque designs in gold and green from wood-blocks by Gregory, Collins & Reynolds.

1845 A Story Book of the Seasons: Spring. Written for young children by Mrs Harriet Myrtle, with illustrations by John Absolon.

Joseph Cundall, 12 Old Bond Street. 165× 125 mm. 152 pp+4 lithographed pl. Printed by C. Whittingham, Chiswick.

1846 A Story Book of the Seasons: Summer. Written for young children by Mrs Harriet Myrtle with illustrations by John Absolon.

Joseph Cundall, 12 Old Bond Street. 165× 125 mm. viii+136 pp+4 lithographs printed in black and 1 tint, signed 'Whittingham, 20 Tooks Ct.' Text printed by C. Whittingham, Chiswick.

The same book appeared n.d. (presumably later) with title-page: *The Little Foundling and Other Tales. A Story Book for Summer.* Written for young children by Mrs Harriet Myrtle.

Joseph Cundall, Old Bond Street.

This edition is same size, same pagination, with same 4 plates, but has been reset and not printed by Whittingham. The plates are printed in black and ochre and hand-coloured.

1846 A Story Book of the Seasons: Autumn. Little Amy's Birthday and Other Tales. Written for young children by Mrs Harriet Myrtle.

Joseph Cundall, 12 Old Bond Street. 165× 120 mm. iv+132 pp+frontispiece drawn on stone. Text printed by C. Whittingham, Chiswick.

1848 The Man of Snow and other Tales. A Story Book for Winter. Written for young children by Mrs Harriet Myrtle.

Joseph Cundall, 12 Old Bond Street.
Osborne Collection

End of Myrtle Story Books

[1845] †A Booke of Christmas Carols edited by Joseph Cundall. Illuminated borders drawn by J. Brandard.

Joseph Cundall, Old Bond Street (in some (later?) copies, Henry G. Bohn, York Street, Covent Garden), n.d. 191×133 mm. Chromolithographed in 6 colours by M. & N. Hanhart and printed from type by Charles Whittingham. Some copies with Cundall's imprint dated 1846.

This book, the first of Cundall's 'illuminated' gift books, was sold in at least 5 variant decorated bindings:
   (i) & (ii) red and gold (or blue and gold) flock paper;
   (iii) paper on boards, printed in colours, embossed and hand-painted, in the French 'romantic' style;
   (iv) cloth blocked in gold;
   (v) leather blocked in gold.
   Illustrated on pp 10, 11.

1845 A Series of Sixty Etchings illustrative of the History of Reynard the Fox. 56 plates by Albert van Everdingen and 4 by Simon Fokke.

Joseph Cundall, 12 Old Bond Street. 207× 140 mm.

Uniform with this volume was published:
1845 Reynard the Fox translated by S. Naylor.

Longman. 207×140 mm. 252 pp printed in red and black throughout by Richards, 100 St Martin's Lane.

With a title page and ten chapter openings printed in red, blue, green and black. Both volumes (one of plates, the other of text) were

in elaborately designed bindings of blind- and gold-blocked white cloth. Both were presumably planned and designed by Cundall.

STANDARD STORY BOOKS

1845 The Life and Adventures of Robinson Crusoe of York, Mariner, etc. by Daniel Defoe, with 6 lithographed illustrations after T. Stothard, hand-coloured.

Joseph Cundall, 12 Old Bond St, & Thomas Edlin, 37 New Bond Street. 198×134 mm. 620 pp. Printed by Charles Whittingham. 10s 6d, or bound in morocco and coloured, 15s.

1845 The Diverting History of John Gilpin [by Oliver Goldsmith] . . . with ten illustrations by a young artist.

Joseph Cundall, 12 Old Bond Street. 218× 278 mm (landscape). 24 pp+10 leaves printed one side only in black line on cream tone by lithography. All illustrations are signed with monogram JL (John Leighton). These are said to be the first published illustrations by John Leighton (1822-1912), although some of his drawings were published in 1844.

Letterpress text printed by Charles Whittingham at the Chiswick Press. In fancy cloth, 5s.

1845 ⋆Little Bo-Peep and other Stories in Rhyme with 12 illustrations by F.L.M.
2s 6d, coloured 4s 6d.

1845 ⋆The Nursery Sunday Book. Written by a Lady. Intended to be read to young children.

With 4 plates. 2s 6d, coloured 3s 6d.

End of 'Standard Story Books'

STORY BOOKS FOR HOLIDAY HOURS
Each volume illustrated with 4 pictures by Absolon, bound in cloth, 2s 6d, or with plates coloured, 3s 6d. 165×127 mm.

1845 The Little Basket-Maker and other Tales printed by Thomas Harrild.

⋆The Two Doves and other Tales printed in large type with 4 coloured pictures.

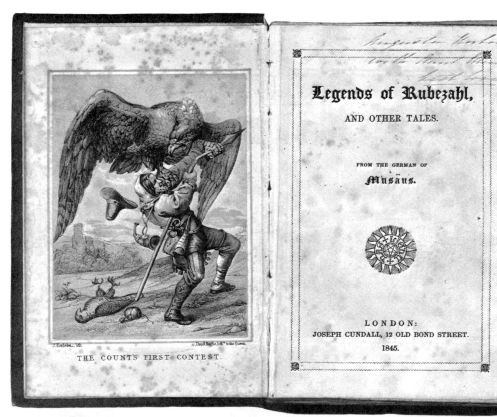

Legends of Rubezahl,

AND OTHER TALES.

FROM THE GERMAN OF

Musäus.

LONDON:
JOSEPH CUNDALL, 12 OLD BOND STREET.
1845.

THE COUNT'S FIRST CONTEST.

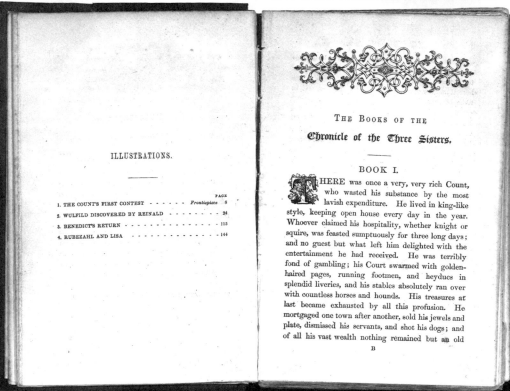

ILLUSTRATIONS.

		PAGE
1. THE COUNT'S FIRST CONTEST	*Frontispiece*	8
2. WULFILD DISCOVERED BY REINALD		24
3. BENEDICT'S RETURN		115
4. RUBEZAHL AND LISA		144

THE BOOKS OF THE

Chronicle of the Three Sisters.

BOOK I.

THERE was once a very, very rich Count, who wasted his substance by the most lavish expenditure. He lived in king-like style, keeping open house every day in the year. Whoever claimed his hospitality, whether knight or squire, was feasted sumptuously for three long days; and no guest but what left him delighted with the entertainment he had received. He was terribly fond of gambling; his Court swarmed with golden-haired pages, running footmen, and heyducs in splendid liveries, and his stables absolutely ran over with countless horses and hounds. His treasures at last became exhausted by all this profusion. He mortgaged one town after another, sold his jewels and plate, dismissed his servants, and shot his dogs; and of all his vast wealth nothing remained but an old

B

Legends of Rubezahl, 1845, in Hazlitt's Holiday Library. Reduced.

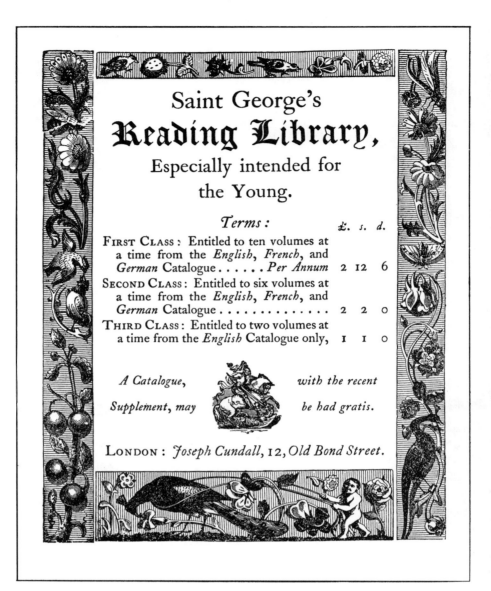

Advertisement from the back of *Peter Parley's Annual*, 1848, set by the Chiswick Press using wood blocks used for Gammer Gurton's Story Books.

1846 The Water Fairy's Gift and other Tales
Printed by John Wilson, Salford, Manchester.

1846 The Miller's Son (Including 'The Miller's Son', 'The Gleaner', and 'The King of the Swans'). 4 hand-coloured lithographs by Absolon.
   Joseph Cundall, 12 Old Bond Street. Printed by John Wilson, Salford, Manchester.

End of 'Story Books for Holiday Hours'

1846 The Two Talismans. From the German of A. L. Grimm, with an illustration by J. B. Sonderland.
   Joseph Cundall. 165× 121 mm. 54 pp in paper covers printed in 2 colours, with same Holbein design as used for Home Treasury. Printed by G. Norman, Covent Garden.

1846 Village Tales from the Black Forest by B. Auerbach. 4 (lithographic?) plates.
   Joseph Cundall, 12 Old Bond Street. 165× 117 mm. Printed by R. & J. E. Taylor, Red Lion Court.

1846 The Value of Money, and other Tales. Written for young children, with illustrations. Contains 'Another Riddle', 'The Washed Doll', 'The Value of Money', 'Naughty Diana', 'The Little Savoyard', 'The Little Marmot', 'The Fire'.
   Joseph Cundall, 12 Old Bond Street. Probably printed by John Wilson, Salford, Manchester.

1846 German Fairy Tales & Popular Stories as told by Gammer Grethel. Translated from the collection of M. M. Grimm by Edgar Taylor, with illustrations from designs by G. Cruikshank and Ludwig Grimm (on wood).
   Joseph Cundall, 12 Old Bond Street. 185× 115 mm. xiv+352 pp+unnumbered titles. Whitehead & Co., Printers, 76 Fleet Street.
   Cruikshank's illustrations to Grimm's *Fairy Tales* were first published in 1823.

1846 Memoirs of a London Doll ed. Mrs Fairstar, with 4 woodcut illustrations after Margaret Gillies.
   Joseph Cundall, 12 Old Bond Street. 194× 133 mm. Printed by Reynell & Weight, Little Pulteney Street.

1846 A Danish Story-Book by Hans Christian Andersen. Translated by Charles Boner. With numerous illustrations by the Count Pocci.
   Joseph Cundall, 12 Old Bond Street. 170× 114 mm. Printed by Robson, Levey & Franklyn, Great New Street, Fetter Lane.
   The Royal Library, Copenhagen

1846 The Good-Natured Bear. With 4 woodcut illustrations after Frederick Tayler.
   Joseph Cundall, 12 Old Bond Street. 197× 127 mm. Printed by F. Skill, 6 Helmet Court, Strand. Price 3s 6d, coloured 4s 6d.

1847 Tales from Denmark by H. C. Andersen. Translated by Charles Boner. Illustrated with 8 hand-coloured lithographs and other illustrations by Count Pocci.
   Joseph Cundall. 12 Old Bond Street. 165× 111 mm. Printed by Levey, Robson & Franklyn.
   1848. Re-issued (in glazed paper on boards with illustration printed on front cover), by Grant & Griffith & Joseph Cundall.

1847 Aunt Carry's Ballads by the Hon. Mrs Norton. Illustrated with 8 hand-coloured lithographs (printed in 2 colours) by John Absolon.
   Joseph Cundall, 12 Old Bond Street. 197× 158 mm. Printed by Charles Whittingham. 5s, coloured 7s 6d.
   Also issued with plates uncoloured, in paper on boards, printed with a beautiful cover design lithographed by M. & N. Hanhart, and lettering on the back.

1847 The Two Flies, a Moral Song. Four hand-coloured plates by (?) J. N. B.
   Joseph Cundall, 12 Old Bond Street. 168× 122 mm. Printed by the Chiswick Press. Issued in red paper wrappers printed with strapwork design in gold.
   Osborne Collection

1847 Hours of Day and Spirits of Night. Illustrations signed: M.E.T. inv. H.F.T. del.
   Joseph Cundall, Old Bond Street. 219× 180 mm. 18 pp lith'd by F. Dangerfield, 26 Bridges St, Covent Garden. Price 6s.
   National Library of Scotland

1847 The Excitement, or A Book to Induce Young People to Read etc., being chiefly extracts from Interesting Books of Travel and Adventure. A New Series. 4 woodcuts after Absolon. Preface dated from Edinburgh, October 1846.

Edinburgh: MacLachlan, Stewart & Co. (London: Joseph Cundall) 139× 88 mm. viii + 388 pp. Printed by Neill & Co. 3s 6d.

This was a revival of a children's annual which had run for at least 10 issues up to 1839, published in Edinburgh by John Johnstone.

1847 The Playmate. A pleasant companion for spare hours.

Joseph Cundall, XII Old Bond Street. 187× 130 mm. viii + 200 pp + 6 etchings on steel. Text printed by G. Barclay.

The Playmate was a monthly periodical for children, which included some material that had already appeared in 'The Home Treasury' and other Cundall publications. Each part consisted of at least 32 pages and 1 etched plate. The title-page was drawn by W. Harvey and engraved on wood by E. Dalziel. The six parts ran from May to October 1847. The volume was issued in paper on boards, with a cover design printed in six colours from wood, signed 'COLLINS & REYNOLDS PRINTERS' and 'R. BURCHETT DEL'. The design incorporated a shield bearing the initials 'JC' (see p 24). A woodcut design pasted down in one of Cundall's family scrapbooks (see p 63) is probably the cover of the parts. A proof of the same design, with the alternative title 'The Schoolfellow' is in the same scrapbook.

1849 The Playmate Second Series. Six further parts from November 1847 to April 1848.

David Bogue, 86 Fleet Street, and J. Cundall, 12 Old Bond Street. iv + pp 203-392 + 7 etchings on steel.

1847 Little Mary's First Going to Church by Lady Charles Fitzroy. Frontispiece and title-page lithographed by Hanhart in (?) 3 colours.

Joseph Cundall, 12 Old Bond St. 152× 120 mm. xii + 228 pp. Printed by Thomas Harrild. Price 4s 6d.

LITTLE MARY'S BOOKS

The prefaces or conclusions to three are signed 'J. C.' and it can be presumed that Cundall planned and edited the series. 'Little Mary' may have been adopted as a title after his daughter Maria (known as 'Maja'), born in 1846.

David Bogue, 86 Fleet Street. 175× 137 mm. Most were printed by G. Barclay, and were illustrated by small wood engravings, by various engravers, of superb quality. In paper covers, price 6d each.

1847 Little Mary's Primer

Little Mary's First Book of Original Poetry

Little Mary's Babes in the Wood by T. Miller.

Little Mary's Scripture Lessons ('Conclusion' signed J. C.)

Little Mary's Reading Book (Introduction signed J. C.)

1849 Little Mary's Picture Book of English History

Little Mary's Spelling Book

1850 Little Mary's Second Book of Original Poetry by T. Miller. Title on title-page: *Original Poems for my Children*. Illustrated by Birket Foster. Printed by Henry Vizetelly.

1865 Little Mary's Treasury of Elementary Knowledge. Adorned with nearly 500 pictures. Containing Little Mary's Primer, Spelling Book, Reading Book, First Book of Poetry, Second Book of Poetry, English History, Scripture Lessons, Babes in the Wood.

Ward & Lock, 158 Fleet Street. 172× 133 mm. Printed by R. Clay, Son & Taylor.

See illustrations on pp 8-9, 51, and 66-69.

End of Little Mary's Books

1847 Elegy written in a Country Churchyard by T. Gray. Illustrated by the Etching Club. Dedicated to the Queen and Prince Albert by the members of the Etching Club.

Published for the Etching Club by J. Cundall, 12 Old Bond Street. 437× 280 mm. 3 preliminary pages + 18 plates from copper printed by Gad & Kenningale all on proof sheets not exceeding 290× 200 mm, pasted down.

The etching Club volumes were published in Large Paper (as this) and Small Paper editions.

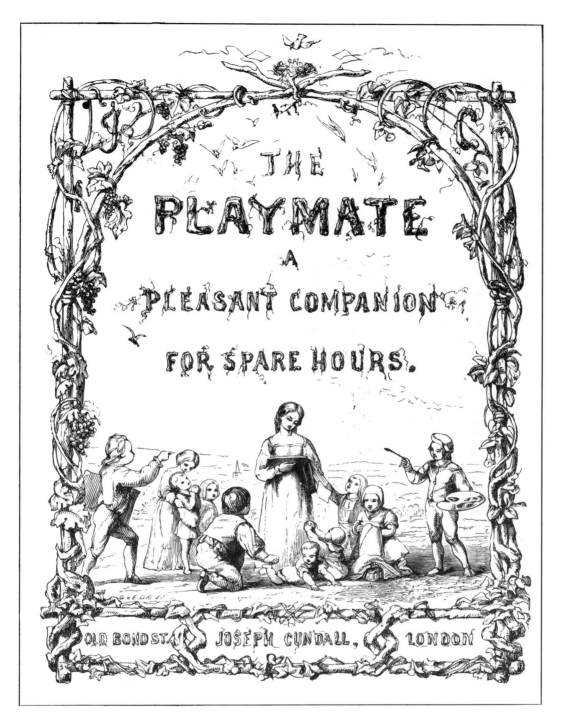

Wood–cut cover design for the *Playmate*, 1847, from a proof in Cundall family scrapbook.

Tales from Denmark, 1847. Woodcut design for cover, from a proof in Cundall family scrapbook.

n.d. (c. 1847) A Book of Stories from the Home Treasury. Illustrated with 30 pictures by eminent artists. Edited by Felix Summerly.

Joseph Cundall, Old Bond Street, n.d. 165× 123 mm. 162 pp+2 pp adverts. Erratically paginated, obviously bound from separate parts. Title-page printed by G. Barclay.

Contents: Jack the Giant-Killer. No imprint. Little Red Riding Hood. No imprint. The Sleeping Beauty in the Wood. Bradbury & Evans. Beauty and the Beast. Bradbury & Evans. Jack and the Beanstalk. G. Barclay. Cinderella. G. Barclay. The More Modern Ballad of Chevy Chase. No imprint. Sir Hornbook. G Barclay. The Sisters; Golden Locks; Grumble & Cheery; The Eagle's Verdict. No imprint.

Two copies examined, with slight differences, both had Cundall's imprint. Listed in British Catalogue of Books as published by Chapman & Hall, Nov. 1846, 7s 6d.

(See entry for *The Home Treasury of Old Story Books*, 1859, p. 83)

1848 The Poetic Prism ed. by Robert Northmore Greville.

Joseph Cundall (Also published by Mac-Lachlan, Stewart & Co, Edinburgh, 1848.) Post 8vo. 10s 6d.

The binding design is illustrated in Cundall's *On Ornamental Art*, 1848.
National Library of Scotland

1848 Letters to an Undergraduate of Oxford. by the Rev. Charles Clarke.

Joseph Cundall, 12 Old Bond Street. 170× 106 mm. viii+78 pp. Printed by Charles Whittingham with decorated initials and head-pieces.
National Library of Scotland

1848 Miss Simmons's Debut (2nd edn). With Fourteen Sketches showing how extremely well the party went off, and how much pleased everybody ought to have been. Hand-coloured lithographed plates, unsigned. Bound in pink paper on boards printed from type.

Joseph Cundall, 12 Old Bond Street. 216× 178 mm. Printed by Robson, Levey and Franklyn. 5s or 7s 6d coloured.

This may have been intended as a children's book, but the illustrations consist mostly of cruel pictures of plain girls, and their even plainer mothers' attempts to find them partners. It does not seem in Cundall's style and one hopes that he was paid to publish it.

THE FAVOURITE LIBRARY
All issued in paper on boards with design of bryony leaves printed by Reynolds & Collins in three colours from wood on yellow paper on front and back, enclosing titles on front.

1848 Mrs Leicester's School by Charles & Mary Lamb. Frontispiece by J. Absolon.

Grant & Griffith; and Joseph Cundall, 12 Old Bond Street. 136×95 mm. Printed by Levey, Robson & Franklyn. Price 1s.

★The Eskdale Herd-Boy by Lady Stoddart (Mrs Blackford). With an illustration by William Harvey. Price 1s.

Announced as 'to be followed by':

1848 The History of the Robins by Mrs Trimmer. With an illustration by W. Harvey.

Grant & Griffith; and Joseph Cundall, 12 Old Bond Street. 136×95 mm. Printed by Levey, Robson & Franklyn. Price 1s.

1848 Memoirs of Bob, the spotted Terrier. With an illustration by W. H. Weir.

Grant & Griffith; and Joseph Cundall, 12 Old Bond Street. 136×95 mm. 126 pp. Printed by Levey, Robson & Franklyn. Price 1s.

End of The Favourite Library

1848 Village Tales from Alsatia by Alexander Weill. Translated from the German by Sir Alexander Duff Gordon, Bart.

Joseph Cundall, 12 Old Bond Street. 181× 130 mm. 248 pp. Printed by R. & J. E. Taylor, Red Lion Court.

1848 The Pentamerone, or The Story of Stories by Giambattista Basile. Translated from the Neapolitan by John Edward Taylor, with illustrations by George Cruikshank (6 etched pl).

180×125 mm. xvi+404 pp. Printed by R. & J. E. Taylor, Red Lion Court, Fleet Street.

The plates in the first edition were uncoloured. The second edition, in 1850, was issued with coloured plates.

1848 †On Ornamental Art applied to ancient and modern bookbinding by Joseph Cundall.

Published at the House of the Society of Arts, John St, Adelphi, and sold by Joseph Cundall, 12 Old Bond Street, and all booksellers. 252× 204 mm. 16 pp text printed by Charles Whittingham with 7 chromolitho and 1 monochrome litho plates by Hanhart and 12 plates printed from wood. £1 1s.

1848 †Words of Truth and Wisdom

Joseph Cundall. 157× 127 mm. 12 leaves of card printed on one side only in chromolithography by F. Dangerfield, Covent Garden. Price 5s.

An 'illuminated' gift book: no name of artist.

1848 †The Creed, the Lord's Prayer & the Ten Commandments

Joseph Cundall, 12 Old Bond Street and David Bogue, 86 Fleet Street. 130× 112 mm. 12 leaves (+ 1 letterpress) printed on one side only in chromolithography by F. Dangerfield, Covent Garden. Price 5s.

An 'illuminated' gift book: name of artist not given. See illustration on p 19.

1848 Grand Historical Pictures. Title-page and 24 full-page line pl printed by lithography one side of paper only, without text. Drawings unsigned, perhaps by Kenny Meadows.

Joseph Cundall, 12 Old Bond Street, and David Bogue, 86 Fleet Street, n.d. Landscape, 185× 280 mm. Paper on board cover with exact copy of title-page on front.

Another copy has second colour, 'sand', added as background to each plate, and dark green border added to cover.

1848 †The Heroic Life and Exploits of Siegfried the Dragon Slayer. With 8 illustrations designed by Wilhelm Kaulbach (lithographed in black and a tone).

Joseph Cundall, Art-Publisher, 12 Old Bond Street; and David Bogue, 86 Fleet Street. 216× 168 mm. Printed by Charles Whittingham. 7s 6d.

The book was issued in gold-blocked cloth and in decorated paper boards with a fine design by R. Burchett, printed in chromolithography by F. Dangerfield. There were a few copies with the title-page printed in red and black and the plates fully hand-coloured. *See illustrations on pp. 17, 25*

1848 The Moon's Histories. Dedication: TO/LORD ALBERT LEVESON GOWER,/THIS LITTLE VOLUME IS INSCRIBED/By the Author,/AS A TESTIMONY OF RESPECT AND GRATITUDE/TO HER GRACE/THE DUCHESS OF SUTHERLAND. According to the Preface, only two of the 18 chapters are translations from the Danish of H. C. Andersen. Name of author not given. With an illustration by John Absolon.

Joseph Cundall, 12 Old Bond Street. 172× 124 mm. xii+ 218 pp. Printed by Thomas Harrild, Silver Street, Falcon Square. 3s 6d.

1849 Nut-Cracker and Sugar-Dolly. Translated by Charles A. Dana, with woodcuts after Lewis Richter.

Joseph Cundall, 12 Old Bond Street; R. Yorke Clarke & Co., Gracechurch Street. 165× 128 mm. Printed by F. A. Brockhaus, Leipzig. 4s.

The book has a cancel title-page printed in England. Probably the main edition was printed (in Germany) for the American market.

A Man Ploughing.

A Farmer sowing Seed.

1849 †Songs, Madrigals and Sonnets.
Preface signed J. C., Camden Cottages,
December 1848.
Longman, Brown, Green & Co. 137×102
mm. 72 pp of which 64 were printed in
colours from wood. 10s 6d.

Messrs Longman's accounts for this work
survive and show that 2,000 copies were
printed and that the colour blocks were
engraved by Collins & Co, who also printed
signatures B and C: signatures D and E and all
the text was printed by Charles Whittingham.
Absolon was paid £36 for 'drawings & designs'.
One-quarter share in the profits was to be paid
to Joseph Cundall; but in January 1851, 40
bound copies and 1,642 copies in sheets were
'remaindered' to H. G. Bohn. See pp 22, 23.

**1849 Illustrations of the Natural Orders of
Plants** by Elizabeth Twining. (Vol. I.)
Joseph Cundall. (Vol. II Day & Son, 1855.)

1849 L'Allegro by John Milton. Illustrated by
the Etching Club.
Joseph Cundall, 12 Old Bond Street. 422×
280 mm. 3 preliminary pages printed letter-
press+20 plates with illustrations printed from
copper and text printed letterpress, on proof
sheets not exceeding 300×200 mm, pasted
down,+2 pp letterpress. The etchings printed
by Gad and Keningale; the letterpress by
Charles Whittingham, Chiswick, in red and
black throughout.

50 copies only of the '4to Colombier' edition
(this) were published at 6 guineas each. 100
copies of the smaller edition (294×210 mm)
bound by Hayday were advertised at 4½
guineas each, and 150 copies in boards at 3
guineas each.

1849 †The Babes in the Wood. Illustrated by
the Marchioness of Waterford (contemporary
advertisement). 10 leaves of card printed on
one side only in chromolithography by M. &
N. Hanhart, lithographed by John Brandard.
Joseph Cundall. 218×141 mm. (Another
copy 235×159 mm.)

1849 †The Babes in the Wood. The same
illustrations, completely re-drawn in larger size,
were also published in 1849 by Joseph Cundall,
etched on copper, elaborately hand-coloured.
The style is not that of a children's book.

Twelve pages on card, 328×235 mm. No
letterpress text. Imprint on etched title-page
illustration: London: Joseph Cundall:
Mdcccxlix.

1851 The Babes in the Wood. New edition
with letterpress title-page on card: *The Babes
in the Wood*; illustrated with ten coloured
drawings by a lady. Printed in colours by
M. & N. Hanhart. New Edition.
Cundall & Addey, 21 Old Bond Street.
222×155 mm. Lithographed in about 8 colours
and gold. 21s.

1861 The same illustrations were also engraved
on wood by W. Dickes and printed in colours
in an edition published by Sampson Low in
1861, uniform with the 'Illustrated Present
Books' edited by Cundall (see p. 80).

1849 The Boy's Almanac for 1849
Joseph Cundall, 12 Old Bond Street, and
David Bogue, 86 Fleet Street. 190 mm×142
mm. Printed by G. Barclay, Castle Street,
Leicester Square.
Osborne Collection

From
*Little
Mary's
Primer*,
1847.

They wash Sheep before they cut off their Wool.

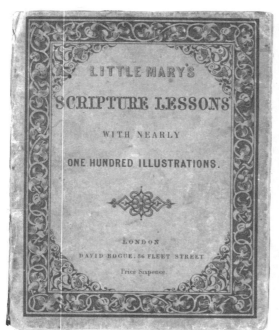

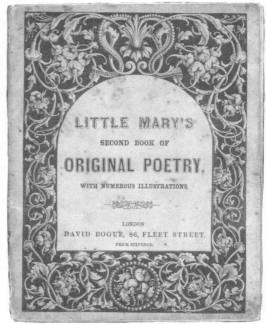

Little Mary's Scripture Lessons, c. 1847. Cover design lithographed in dark brown ink on brown paper. Reduced.

Little Mary's Second Book of Original Poetry, 1850. Cover design printed from wood in terra-cotta on cream paper. Reduced.

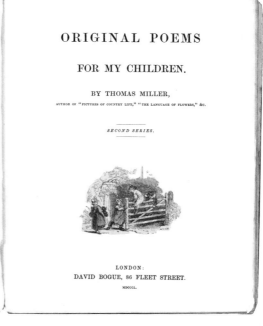

Title-page opening of *Little Mary's . . . Original Poetry* above, with illustrations cut on wood after Birket Foster. Reduced.

1849 †The Peacock at Home by Mrs Dorset (sister of the late Mrs Charlotte Smith). Illustrated and illuminated by Her Grand Niece Mrs W. Warde.

Joseph Cundall, 21 Old Bond Street. 224× 174 mm. 12 pages of stiff card printed on one side only. Type obviously printed by Charles Whittingham (but not stated to be). Borders hand-coloured on lithographed outline. 10s 6d.

Previously published by N. Hailes. See below, in 1851, for another edition of *The Peacock at Home*.

Illustrated on pp 28, 73.

1849-1850 Harry's Ladder to Learning

D. Bogue, 86 Fleet Street, and Joseph Cundall, 21 Old Bond Street. 141× 112 mm. 240 pp with 230 wood-engraved illustrations. Printed by G. Barclay.

Reissued by David Bogue, 86 Fleet Street, n.d.

Containing six parts, which were also issued separately:

1. *Harry's Horn-Book*, pp 1-64

2. *Harry's Picture-Book*, pp 65-112

3. *Harry's Nursery Songs*, pp 113-144

4. *Harry's Nursery Tales*, p 145-174

5. *Harry's Simple Stories*, pp 175-208

6. *Harry's Country Walk*, pp 209-238

'Harry' was Cundall's second child and eldest son, born 1847.

From *Little Mary's . . . Original Poetry*, 1850.

1850 A Treasury of Pleasure Books for Young Children with more than 100 illustrations by John Absolon and Harrison Weir.

Grant and Griffith, successors to Newbery and Harris, St Paul's Church Yard; and Joseph Cundall, Old Bond Street. 172× 111 mm. viii +280 unnumbered pp. Printed by G. Barclay. Set in Caslon Old Face. Plain 3s 6d, coloured 6s.

The foreword 'To my dear children, Maja, Harry, and Herbert' is signed 'J.C., Kentish Town, 1849'. Among other things he recommends his readers to copy the patterns of the designs for the cover and end-papers by Owen Jones, 'which I think are very pretty'.

This book and its illustrations were reprinted many times in various forms. At least 18 of the stories were issued singly as 'Pleasure-Books for Young Children', price 6d each plain, one Shilling coloured, and were so advertised by Cundall & Addey, 21 Old Bond Street, in 1851. They were also re-issued as *Pleasure Books for Children* by Sampson Low. See cover designs illustrated in *Victorian Book Design*, 2nd edn., 1972, pp. 62-63.

See *A Treasury of Pleasure Books for Young and Old*, 1851, p. 71, and *A Treasury of Pleasure Books for Young People*, p. 84.

1850 The Book of Ruth. Dedicated to H.R.H. the Duchess of Cambridge. 'Published for the Benefit of Charitable Institutions in the Parish of Lower Chelsea.' Illustrated by The Lady Augusta Cadogan.

Joseph Cundall, 21 Old Bond Street. 233× 178 mm. Printed by C. Whittingham, Chiswick. 24 pp text set in black letter+ 8 plates etched (on copper?) in line and very richly hand-coloured. Area of illustration: 165× 115 mm.

Copy examined was bound in blue-green cloth lettered in gold on spine only, probably not as issued. The book was probably issued in lithographed pictorial paper covers on boards, with some copies in leather, as for example, *Siegfried the Dragon-Slayer* and other books issued by Cundall.

1850 A Glance at the Exhibition of the Royal Academy 1850 (attrib. by W. M. Rossetti to Dr Waagen, *Pre-raphaelite Diaries and Letters*, London 1900, p 274, n.).

Joseph Cundall, 21 Old Bond Street. 276× 203 mm. Printed by G. Barclay, Castle Street, Leicester Sq.

1850 Illustrations to Southey's Roderick. Designed and drawn on stone by F. B. N. (Miss Newdegate).

Joseph Cundall, Art Publisher, 21 Old Bond Street. 360× 270 mm. 'To be completed in 8 parts, each 7s 6d.' Part 2 contains 9 plates lithographed in black and cream on l.h. pages, with letterpress text facing.
National Library of Scotland

1850 Rip Van Winkle by Washington Irving. Illustrated with six etchings on steel, by Charles Simms, from drawings by Felix Darley (New York), reduced by Daguerrotype and hand-coloured.

Cundall, 21 Old Bond Street. 187× 127 mm. 32 pp+plates. Printed by W. M'Dowall, Little Queen Street, Lincoln's-Inn Fields. 5s.

'The Present Illustrations have been reduced from the originals, which are much larger, by the agency of the Daguerrotype, and I hope that the expression of every line has been most faithfully preserved . . .
Joseph Cundall, April 1850.'

Issued in white glazed paper on boards, printed from type in blue and red.

1850 The Great Wonders of Art by Arthur C. Wigan.

Published for Joseph Cundall by Sampson Low & Son, 47 Ludgate Hill. 185× 133 mm. Printed by Thomas Harrild, Silver Street, Falcon Square, London.
See 1856, Books for Young Readers.

1851 †The Peacock at Home by Mrs Dorset (Sister of the late Mrs C. Smith). Illustrated and illuminated.

Cundall & Addey, Old Bond Street. 150× 117 mm. 24 pp printed on right-hand pages by chromolithography (in red, blue, green, gold and brown) and letterpress (for type), on left-hand pages: type in black and woodcut borders

in red. Verso of title-page bears imprint 'Printed by C. Whittingham 1851' and Chiswick Press shield. No name of lithographic printer: advertisement of 1851 states 'Illuminated borders printed by M. & N. Hanhart. Extra boards, gilt edges, price 5s'. Book is in the same style as 1849 edition with illustrative coloured borders round type, but the artwork is quite different.

1851 †The Story of Jack and the Giants illustrated with 35 woodcuts after drawings by Richard Doyle, engraved by G. & E. Dalziel.

Cundall & Addey, 21 Old Bond Street. 210 × 152 mm. Printed by Robson, Levey & Franklyn. 'Handsomely bound, price 2s 6d or in cloth 3s 6d; with Coloured Plates, gilt edges, 6s.'

The illustrations were commissioned by the Dalziels, who presumably took the financial risks of publication.

The book was republished in 1858 by Griffith & Farran, printed by the same printers, with the 8 full-page illustrations coloured by hand.

See *Victorian Book Design*, 2nd edn, 1972, pp 142, 144.

[c.1851] A Laughter Book for Little Folk. Translated from the German by Mme de Chatelain. Illustrated by Theodor Hosemann.

Cundall & Addey, 21 Old Bond Street and William Tegg & Co, 85 Queen Street, Cheapside (n.d.). 238× 185 mm. pp 1-20 printed one side only in line from stone, hand-coloured. 'Twenty large comic coloured drawings.' Price 2s 6d.

Second edn inscribed 1852. On p 20: W. M'Dowall, printer, Little Queen Street, Lincoln's-Inn-Fields.

1851 The Pleasures of the Country. Simple Stories for Young People by Mrs Harriet Myrtle with eight illustrations (hand-coloured wood-cuts) by John Gilbert.

Cundall & Addey, 21 Old Bond Street. 210× 159 mm. iv+116 pp+8 plates. Printed by W. M'Dowall, Little Queen Street, Lincoln's-Inn-Fields.

1851 Magic Words: A Tale for Christmas Time by Emilie Maceroni, with 4 illustrations by E. H. Wehnert, lithographed in colour by Day & Son.

Cundall & Addey, 21 Old Bond Street. 160× 100 mm. 56 pp.

Osborne Collection

1851 Merry Tales for Little Folk. Edited by Madame de Chatelain. Illustrated with more than 200 pictures (woodcuts).

Cundall & Addey, 21 Old Bond Street. 133× 93 mm. vi+452 pp+4 pp. Title printed from wood in 2 colours. Printed by W. McDowall, Lincoln's-Inn-Fields. Plain 3s 6d, coloured 4s 6d.

A Cundall & Addey advertisement states: 'This volume contains about forty of the long-established favourite Stories of the Nursery in England and abroad, re-written or re-translated from the original Authors, by MADAME DE CHATELAIN. In performing her labour of love, the Editor has adhered, as far as possible, to the words of the Authors, and has never attempted to improve the old familiar Rhymes, her only aim having been to weed out such vulgarisms as modern taste reproves.'

Reissued by Crosby, Lockwood & Co., 7 Stationers' Hall Court, Ludgate Hill, n.d. (copy inscribed 1884). Binding in cloth blocked in gold with title 'Forty Favourite Fairy Tales'.

1851 †The Church Catechism. Illuminated with border designs from an ancient Italian missal.

Cundall & Addey. Old Bond Street. 133× 108 mm. Illuminated title page and 10 leaves printed one side only. Colour plates signed: 'Leighton Brothers Lith'. Text printed by C. Whittingham, Chiswick. 6s.

An 'illuminated' gift book.

1851 †A Chaplet of Pearls. Rhymes and Fragments of Ancient & Modern Verse illustrated & lithographed by Mrs Charles Randolph.

Cundall & Addey, 21 Old Bond Street. 273× 181 mm. 24 pages of card printed one side only by chromolithography. Text printed letterpress probably by C. Whittingham.

A secular 'illuminated' gift book.

1851 The Poetical Works of Oliver Goldsmith. With 30 illustrations by John Absolon, Birket Foster, James Godwin and Harrison Weir, engraved on wood.

Cundall & Addey, 21 Old Bond Street. 178× 115 mm. xvi+134 pp+2 pp advts. Printed by Thompson & Davidson, Great St Helens.

1851 A Treasury of Pleasure Books for Young and Old. With 36 illustrations by Edward Wehnert and Harrison Weir. New Series.

Cundall & Addey, 21 Old Bond Street. Cambridge, U.S.: J. Bartlett. 173× 113 mm. 144 pp. Printed by Robson, Levey & Franklyn, Great New Street, London. Cloth, 3s 6d; coloured, gilt edges, 6d.

Foreword is addressed: 'To my dear children, Maja, Harry and Herbert' and signed 'J.C., Kentish Town, Oct. 1851'. There are 6 full-page illustrations for each story: The Charmed Fawn, Robin Hood, Ugly Little Duck, Puss in Boots, Hans in Luck, and Peter the Goatherd.

Courtesy Justin G. Schiller, Ltd.

See entry for *A Treasury of Pleasure Books for Young Children*, 1850.

Variant monograms designed by W. Harry Rogers for Joseph Cundall.

1851 The Comical Creatures from Wurtemberg. With 20 illustrations drawn from the stuffed animals contributed by Hermann Ploucquet of Stuttgart to the Great Exhibition. The illustrations are woodcuts from drawings by Harrison Weir, hand-coloured.

David Bogue, Fleet Street. 206×160 mm. 96 pp.

1851. Second edn.

This book was in a collection, belonging to Joseph Cundall, of books published, designed or written by him, but may have nothing to do with him.

1851 Home Pictures. Sixteen Domestic Scenes of Childhood. Drawn and etched by Hablot Knight Browne.

Cundall & Addey, 21 Old Bond Street. 262×190 mm. 19 leaves of card printed one side only (except t.p.) of which 17 are circular etchings, hand-coloured. No text. Imprint of Robson, Levey & Franklyn. 12s plain, 21s coloured.

See also entry for *A Day of Pleasure*, 1853.

1851 Animals from the Sketch Book of Harrison Weir

Cundall & Addey, 21 Old Bond Street. 276×212 mm. 24 plates printed in black from wood blocks on buff flat tint, by Robson, Levey & Franklyn. 'Cloth, price 7s 6d; or in Six Parts at 1s each; Coloured after the Artist's Original Drawing, cloth gilt edges, 31s 6d.'

1851 *Home for the Holidays. New edition, with nine illustrations by Kenny Meadows. 4to in Ornamental Cover, price 1s. (Cundall & Addey advert in 1851.)

1851 †Choice Examples of Art Workmanship selected from the Exhibition of Ancient & Mediaeval Art at the Society of Arts. Drawn and engraved under the superintendance of Philip De la Motte.

Cundall & Addey, 21 Old Bond Street; George Bell, 186 Fleet Street. 280×184 mm. 60 monochrome wood engravings+title+14 pp. text. Printed by Robson, Levey & Franklyn, Great New Street. Also large paper, 324×241 mm, with all plates hand-coloured. 'A few vellum copies are printed. These are most carefully illuminated and finished by Mr De la

Motte, bound in velvet, price Twelve Guineas.' The ordinary edition, Imp 8vo, was 25s 'elegantly bound with gilt bosses in facsimile of an ancient Venetian Binding'. Coloured morocco was £4 4s. Large paper £3 3s, coloured £6 6s.

See illustration on p 27.

1852 New Tales from Faëry Land. Illustrated with 4 unsigned wood-engravings in black on yellow tint.

Addey & Co. (Late Cundall & Addey) 21 Old Bond Street. 178×120 mm. Printed by Petter, Duff & Co.

1852 The Village Queen by Thomas Miller. With 4 illustrations by E. Wehnert, J. Absolon, William Lee & H. Weir.

Addey & Co. (late Cundall & Addey). 280×222 mm. 108 pp+4 plates printed from wood in 6 to 8 colours, pasted down on blank pages. Text printed by G. Barclay. Colour plates printed by Leighton Bros.

1853. Second edn published by Addey & Co., 21 Old Bond Street.

This book was exhibited on Cundall's stall at the Great Exhibition in 1851.

1852 The Little Sister by Harriet Myrtle, with 16 illustrations by H. J. Schneider.

Addey & Co., 21 Old Bond Street (Late Cundall & Addey). 162×213 mm. landscape. Text printed by C. Whittingham. 16 copper plates coloured by hand.

1852 Stories Told in Pictures. A morning Ride mid Country Scenes.

George Routledge & Co., Farringdon Street. 140×215 mm. Large picture, mounted on cloth. Bound in green cloth blocked in gold. No mention of J. C., but was among his own books and was probably planned by him.

Advt. at back: Nearly ready: A Noon-Day Ramble, etc.; a coloured picture six feet in length. An Evening Walk, etc.

1853 A Day of Pleasure by Mrs Harriet Myrtle, with 8 illustrations by Hablot K. Browne.

Addey & Co., 21 Old Bond Street.

Contains 8 of the etchings from *Home Pictures* (1851) trimmed to octagonal shape, uncoloured. Sq. 8vo, iv+104 pp.

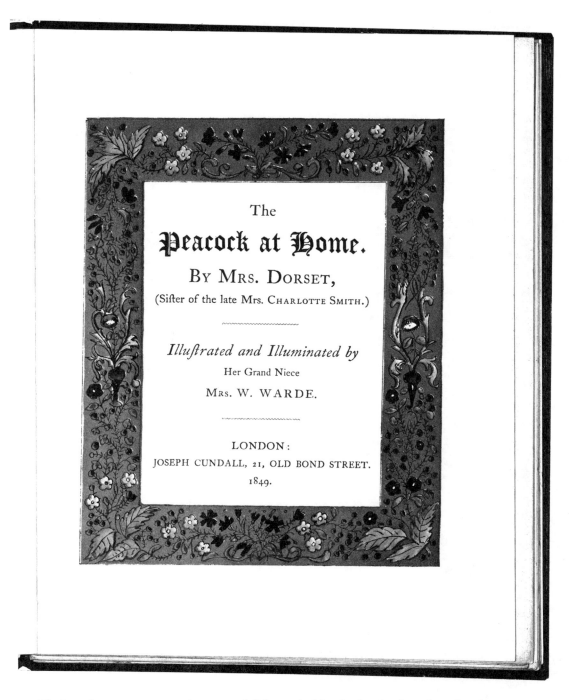

The
Peacock at Home.

By Mrs. Dorset,
(Sister of the late Mrs. Charlotte Smith.)

Illustrated and Illuminated by
Her Grand Niece
Mrs. W. Warde.

LONDON:
JOSEPH CUNDALL, 21, OLD BOND STREET.
1849.

The Peacock at Home, 1849. Title-page with lithographed border, hand-coloured. Reduced.

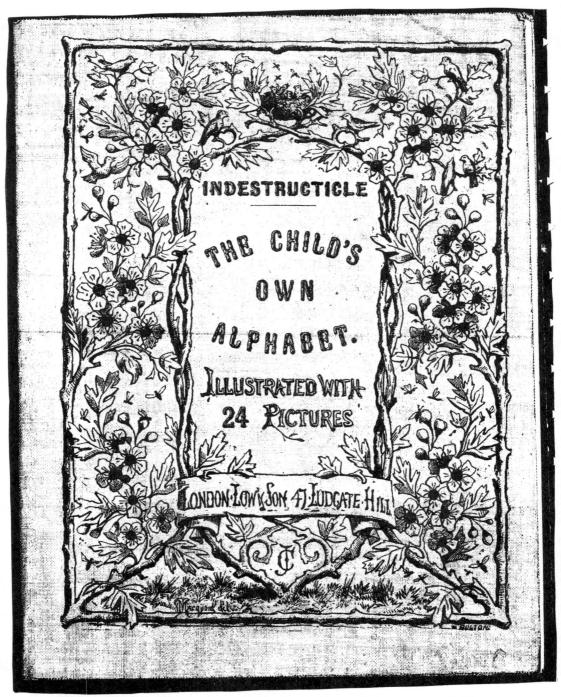

The Child's Own Alphabet, c. 1852. Cover design by T. Macquoid, incorporating 'J C' monogram, printed from wood on pink cloth. Reduced.

ADDEY & CO.

Cundall went bankrupt, possibly in 1849, and in an entry of 1849 in the Chiswick Press account books, Addey is also entered as a bankrupt. Cundall's partnership with Addey began in 1849 and seems to have terminated in 1852 when Cundall moved to 168 New Bond Street. Addey remained at 21 Old Bond Street, at least until 1854. By 1856 he was at Henrietta Street, Covent Garden; the latest date on books with 'Addey & Co.' so far found is 1857.

Books originally published by Cundall were republished by Addey & Co., and Addey published books written or edited by Cundall after 1852. Other books published by Addey look like 'Cundall books', and it may be that Cundall continued to work behind the scenes with or for Addey, but this is a guess. It is at least probable that the early books published by Addey were worked on, if not started, by Cundall.

1852 The Picture Pleasure Book

illustrated with upwards of 500 engravings from drawings by eminent artists.

Addey & Co., 21 Old Bond Street, n.d. 340× 275 mm. 64 pp (iv+ 1-60 numbered pages +2 pp advt dated Dec. 1852) of wood blocks from miscellaneous children's books, e.g. *A Treasury of Pleasure Books*, already published by Cundall & Addey, with short captions or couplets under some of the blocks. Printed by G. Barclay. Bound in paper on boards with pictorial design by John Leighton, cut on wood by H. Leighton, printed in red and black on eggshell blue paper. 6s. Originally published in monthly parts at 6d, containing 50-60 engravings 'on stout paper, and stitched in a wrapper'.

See illustration on p. 76.
1853. Second series.

1852 Aunt Effie's Rhymes for little children

by Ann, Lady Hawkshaw. With 24 illustrations by Hablot K. Browne (cut on wood).

Addey & Co., 21 Old Bond Street (late Cundall & Addey). 184× 133 mm. 96 pp. Printed by Petter, Duff & Co., Crane Court, Fleet Street. Republished 1878 by Routledge.

PATENT INDESTRUCTIBLE PLEASURE BOOKS.

This series, printed on cloth and sold at 1s, may have been based on the 'Treasury of Pleasure Books', and was published by Cundall & Addey 1851-52, Addey & Co., and Low & Son (later Sampson, Low), 47 Ludgate Hill.

c. 1852 Mother Goose and Simple Simon.

Edited by Joseph Cundall. Illustrated by John Absolon. 178× 133 mm.

c. 1852 The Child's Own Alphabet

184× 138 mm. Illustrated with 24 pictures by Absolon. Cover printed from wood on pink limp cloth, designed by T. Macquoid, incorporating 'J.C.' monogram.
See illustration on p 74.

c. 1852 Maja's Alphabet was a coloured edition

of the above, printed on paper, gilt edges, price 1s.

Cundall & Addey; and David Bogue.

Also announced at one shilling each:

The Indestructible Primer

The Indestructible Spelling Book

The Indestructible Reading Book

The Indestructible Lesson Book (Four Parts

bound together in 1 volume, 90 pictures, priced at 5s.)

End of Patent Indestructible Books

1853 The Charm. A Book for Boys and

Girls. Illustrated with monochrome woodcuts.
Foreword signed 'J. C.'.

Addey & Co., 21 Old Bond Street. Edinburgh: J. Menzies. Dublin: J. McGlashan. 194× 133 mm. Printed by G. Barclay.

This was the publication in Annual form of a sixpenny monthly periodical edited by Cundall from 1852 to 1854.

1854: *The Charm*, second series.
1855: *The Charm*, third series. Both printed by G. Barclay.

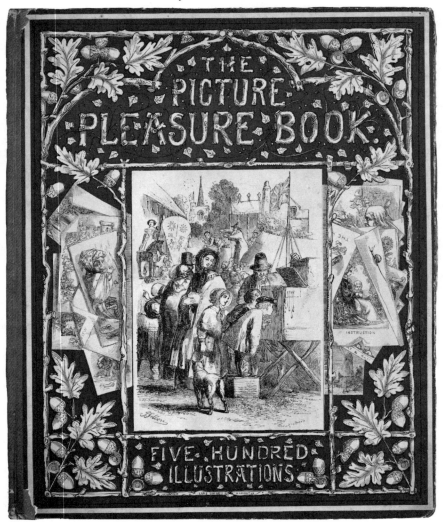

The Picture Pleasure Book, 1852. Cover design by Luke Limner (John Leighton) cut on wood by H. Leighton, and printed in red on green paper boards. Reduced.

1853 The Diary and Houres of the Ladye Adolie, a faythfulle Childe.

Addey & Co., 21 Old Bond Street. 216× 158 mm. Printed by Charles Whittingham in the style of *The Diary of Lady Willoughby*. A fine example of Chiswick Press typography and almost certainly a book associated with Cundall.

1853 Child's Play. 17 drawings by E. V. B.

Addey & Co., Old Bond Street. 248× 165 mm. Printed at Appel's Anastatic Press. 8s, or 31s 6d with coloured plates.

The text is incorporated in the drawings: there is no type-setting in the entire book, as the title-page is also drawn.
See also under 1859.

1853 The Poetical Works of Edgar Allan Poe, edited by James Hannay, with 20 illustrations by E. H. Wehnert, James Godwin, F. W. Hulme, & Harrison Weir.

Addey & Co., 21 Old Bond Street. 175× 108 mm. xxxii+ 144 pp+advts. Ptd by G. Barclay.
See 16 pp of adverts of Addey books at back.

76

1853 Children's books from Addey & Co's list, c. 1853, *not seen*:

★The Charm Almanack embellished with twenty large engravings. 8vo, sixpence.

★Grimm's Household Stories, 2 vols, embellished with 200 small and 36 full-page illustrations by E. H. Wehnert. Cr. 8vo, cloth gilt, 12s. Also in 9 parts, 1s each.

★A Life of Wellington for Boys by W. K. Kelly, with 8 illustrations by John Absolon. F'cap 8vo, 6s cloth gilt.

★The Green Bird. 4to, 5s cloth; coloured plates, 7s 6d.

★A Hero. Philip's Book; a Tale for Young People. Illustrated by James Godwin. F'cap 8vo, 3s 6d cloth gilt.

★Arbell. A Tale for Young People by Jane Winnard Hooper, illustrated by James Godwin. F'cap 8vo, 6s cloth gilt.

★The Little Drummer by Gustav Nieritz, translated by H. W. Dulcken, illustrated by John Gilbert. F'cap 8vo, 2s 6d cloth gilt.

★Naughty Boys and Girls. Comic Tales and Coloured Pictures from the German of Dr Julius Bahr, by Madame de Chatelain. Post 4to, in coloured picture binding, 2s 6d.

★Wonder Castle. A Structure of Seven Stories by A. F. Frere, illustrated by E. H. Wehnert. Sm. 4to, cloth gilt, 6s.

★The Apple Dumpling. Handsomely bound with coloured pictures, 3s 6d.

★An Ancient History abridged from Rollin, for the use of Young People by Mary G. Wilkes. 2nd ed, 4 illustr. by E. H. Wehnert. 18mo, cloth, 3s 6d.

★Cinderella. 13 illustr. by M. J. R. 4to, extra boards, 5s or with coloured pictures 7s 6d.

Kindness and Cruelty or the Adventures of Edward and Stephen. 4 illustr. 16mo, 1s.

★Wild Spring Flowers by Alice Georgina, aged eight years. 4to in wrappers, 1s.

★Golden Songs for Silvery Singers. Printed in gold on coloured paper. In oblong 16mo, 1s.

BERTIE'S INDESTRUCTIBLE BOOKS
Printed on cloth expressly prepared, 6d each.

No. 1 **Bertie's Horn Book**

No. 2 **Bertie's Word Book**

No. 3 **Bertie's Farm Yard**

No. 4 **Bertie's Woodside**

No. 5 **Bertie's Wild Beasts**

No. 6 **Bertie's Foreign Birds**
   'Bertie' was Joseph Cundall's third child and second son, b. 1848.

ADDEY & CO. LITTLE FOLK'S BOOKS
All 140 × 102 mm., bound in green paper on boards printed with design in red, one shilling each.

★No. 1 Funny Rhymes and Favourite Tales

★No. 2 Nursery Heroes

★No. 3 Nursery Heroines

★No. 4 Fairy Folk and Wonderful Men

1852 No. 5 **Far-Famed Tales.** From the Arabian Night's Entertainments. Illustrated with 40 engravings.
   Addey & Co., 21 Old Bond Street, 128 pp. Printed by Petter, Duff & Co., Playhouse Yard, Blackfriars.

1853 No. 6 **Aladdin and the Wonderful Lamp** and **Sinbad the Sailor.** Illustrated with 31 engravings.
   Addey & Co., 21 Old Bond Street. Printed by Petter, Duff & Co.

1852 No. 7 **New Nursery Songs for All Good Children** by Mrs Follen.

Addey & Co., 21 Old Bond Street. viii+96 pp.

*No. 8 **The Little Fortune Teller** by Caroline Gilman. Illustrated by Kenny Meadows.

End of Addey & Co.'s Little Folk's Books

1853 The Practice of Photography. A Manual for Students and Amateurs. By Philip H. De la Motte. Illustrated with a calotype portrait taken by the Collodion Process (pasted down as frontispiece, varying from copy to copy).

Joseph Cundall, 168 New Bond Street. 182×114 mm. viii+150 pp. Printed by G. Barclay, Castle Street, Leicester Square. 4s 6d.

Advertisement at back announces:
Now ready, price 16s.
Photographic Studies by George Shaw.
Preparing for publication, in Parts price 21s each. *The Progress of the Crystal Palace at Sydenham . . .* by Philip De la Motte.
A Series of Photographic Pictures. By Hugh Owen.
A Series of Photographic Pictures of Welsh Scenery by J. D. Llewellyn. Published in parts. 10s 6d each.
The Photographic Album
4 parts, containing photos by Roger Fenton, Philip De la Motte, Alfred Rosling, Hugh Owen & Joseph Cundall.

*1853 **Photographic Studies** by George Shaw.

Joseph Cundall. 2 parts fol., 16s each.
London Catalogue of Books

1853 The Discontented Chickens (or, the History of Gockel and Scratchfoot). Home Story Books. Illustrated with 8 wood-engravings.

David Bogue, 86 Fleet Street. 165×112 mm. viii+94 pp.

No mention of Cundall or proof that it is to do with him, but in Cundall's collection of his own books. An advertisement at the back lists Cundall's 'Indestructible Books'.

Reprinted 1859.

1853 A Children's Summer. Eleven etchings on steel by E. V. B. Illustrated in prose and rhyme by M. L. B. and W. M. C.

Addey & Co., 21 Old Bond Street. 278×368 mm. Oblong. 31 pp+9 etchings (2 etchings appear as vignettes on text pages). Printed by C. Whittingham, Tooks Court, Chancery Lane. No proof of connection with Cundall.
Courtesy Justin G. Schiller Ltd, New York

1853 †The Poets of the Woods. Twelve Pictures of English Song Birds. With 12 colour illustrations from water colour drawings by Joseph Wolf.

Thomas Bosworth, Regent Street. 248×178 mm. vi+58 pp. Text printed by C. Whittingham, plates in chromolithography by M. & N. Hanhart.

Identified as a 'Cundall book' in *Sabbath Bells*, 1856 (q.v.)

Illustrated in *Victorian Book Design*, 2nd edn, 1972, p 147.

1853 †Poetry of the Year. Passages from the Poets descriptive of the Seasons. With 22 coloured illustrations from drawings by eminent artists, including Birket Foster, Harrison Weir, T. Creswick, David Cox, Joseph Wolf and E. V. B.

George Bell, 186 Fleet Street. 276×188 mm. Text printed by G. Barclay, plates in chromolithography by M. & N. Hanhart and Day & Son. The colour plates are pasted down on heavy cartridge text pages.

Identified as planned by Cundall in *Sabbath Bells*, 1856 (q.v.)

[*c.* 1867] Re-issued by Charles Griffin & Co., Stationers Hall, Paternoster Row (n.d.) with a new chromolithographed title-page by Arthur Warren, and 16 chromolithographs probably by Vincent Brooks (who signed the title-page). Text printed by J. & W. Rider, Bartholomew Close.

1854 †Feathered Favourites. Twelve coloured pictures of British Birds from drawings by Joseph Wolf.

Thomas Bosworth, 215 Regent Street. 250×178 mm. Text printed by C. Whittingham. 12 plates in chromolithography exactly like *The Poets of the Woods*, so probably by M. & N. Hanhart.

Uniform with *The Poets of the Woods*, 1853.

1854 A Poetry Book for Children. 20 large illustrations by various artists. Tail-pieces drawn by Harrison Weir and engraved by S. V. Slader. Preface in verse signed J. C.

G. Bell, 186 Fleet Street. 174× 117 mm. 144 pp. Printed by R. Clay.

1854 The Old Story-Teller. Popular German Tales collected by Ludwig Bechstein. One hundred illustrations by Richter.

Addey & Co., 21 Old Bond Street. Printed by Levey, Robson & Franklyn. 8 full page hand-coloured plates.

No mention of J. C., but found among his own books.

1854 Pretty Poll. A Parrot's Own History. 4 hand-coloured illustrations by Harrison Weir.

Addey & Co., 21 Old Bond Street. 170× 128 mm. 84 pp+advts. Printed by L. Thompson, Great St Helen's, London.

1854 Happy Days of Childhood by Amy Meadows. 24 Plates by H. Weir, and hand-coloured frontispiece by Birket Foster.

Cundall & Sampson Low. 203× 158 mm. 52 pp. Printed by G. Barclay.

1854 Little Susy's Six Birthdays by her Aunt Susan (Mrs Elizabeth Prentiss). Illustrated with four engravings.

Published for Joseph Cundall by Sampson Low & Son, 47 Ludgate Hill. 150× 110 mm. iv+ 190 pp.

Osborne Collection

1854 The Water Lily by Harriet Myrtle. Illustrated by H. K. Browne, engraved by T. Bolton.

T. Bosworth, 215 Regent Street. 184× 133 mm. Printed by Petter & Galpin.

No mention of J. C., but found among his own books.

1854 Mia and Charlie by Harriet Myrtle. Illustrated by B. Foster.

D. Bogue. Printed by H. Barclay, Winchester.

Stated by H. M. Cundall to have been produced by his father.

SAMPSON LOW'S 'ILLUSTRATED PRESENT BOOKS'

This title appears above an advertisement for some books in the series. The series was planned by Cundall, and nearly all the titles bear his name as publisher, or his monogram. The eleven titles that follow appeared in a page size 196× 130 mm approx. and were bound in a similar but not identical 'family' style, in varying colours of cloth.

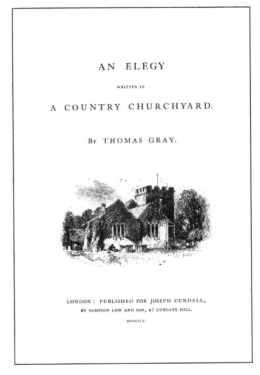

1854 An Elegy written in a Country Churchyard by Thomas Gray. Illustrations by Birket Foster, George Thomas and A Lady (E. V. B.), engraved on wood.

Published for Joseph Cundall by Sampson Low & Son, 47 Ludgate Hill. 24 leaves printed rectos only. Printed by G. Barclay.

1861. Reprinted for Sampson Low by R. Clay.

1855 The Deserted Village by Oliver Goldsmith. Illustrated by the Etching Club.

Published for Joseph Cundall by Sampson Low & Son, 47 Ludgate Hill. Printed by R. Clay.

The original etchings were copied onto wood.

1855 The Pleasures of Hope by Thomas
Campbell. Illustrated by Birket Foster, George
Thomas and Harrison Weir. J. C.'s monogram
on verso of title-page.

Sampson Low & Son, 47 Ludgate Hill.
Printed by R. Clay.

1861. Reprinted by Clay for Sampson Low
with J. C.'s monogram.

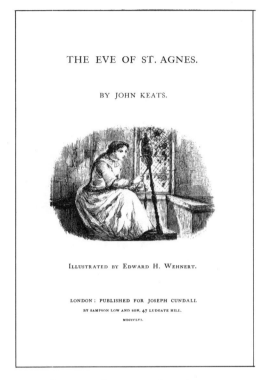

THE EVE OF ST. AGNES.

BY JOHN KEATS.

ILLUSTRATED BY EDWARD H. WEHNERT.

LONDON: PUBLISHED FOR JOSEPH CUNDALL
BY SAMPSON LOW AND SON, 47 LUDGATE HILL.
MDCCCLVI.

1856 The Eve of St Agnes by John Keats.
Illustrated by E. H. Wehnert. J. C.'s monogram on verso of title-page.

Published for Joseph Cundall by Sampson
Low & Son, 47 Ludgate Hill. Printed by R.
Clay.

1859: Reprinted by R. Clay for Sampson
Low, with J. C.'s monogram on verso of title-
page.

**1856 Songs of the Brave. The Soldier's
Dream, and other Poems and Odes.**
Illustrated by B. Foster, E. Duncan, G. Thomas,
A. Huttula. J. C.'s monogram on title-page.

Sampson Low, Son & Co., 47 Ludgate Hill.
No printer mentioned.

1857 The Rime of the Ancient Mariner by
S. T. Coleridge. Illustrated by E. H. Wehnert,
B. Foster, E. Duncan. J. C.'s monogram on
verso of title-page.

Sampson Low, Son & Co., 47 Ludgate Hill.
Printed by R. Clay.

1857 The Farmer's Boy by Robert Bloom-
field. Illustrated by B. Foster, H. Weir, G. E.
Hicks.

Sampson Low. Printed by R. Clay.

No mention of J. C., but stated by H. M.
Cundall to have been supervised by him.

1858 Pastoral Poems by W. Wordsworth.
Illustrated by B. Foster, G. Thomas, H.
Warren. J. C.'s monogram on verso of title-
page.

Sampson Low, Son & Co., 47 Ludgate Hill.
Printed by Edmund Evans.

1859: Reprinted with J. C.'s monogram on
verso of title-page.

1859 The Hamlet by Thomas Warton.
Illustrated with 14 etchings by Birket Foster.

Sampson Low, Son & Co., 47 Ludgate Hill.

The illustrations are printed from copper, the
only book with copperplates in this series. No
mention of Cundall, but in identical format.

1861 The Babes in the Wood. Illustrated
with colour plates printed from wood by W.
Dickes.

Sampson Low, Son & Co.

No mention of J. C. (see p 67, 1849.)

1863 Shakespere's [sic] **Songs and Sonnets.**
Illustrated by John Gilbert.

Sampson Low, Son & Co., 47 Ludgate Hill.
195 × 130 mm. 56pp. Printed by Edmund Evans.

The wood engravings are the same as those
in *Shakespeare's Songs and Sonnets*, Sampson
Low, 1862, of which Cundall was editor (q.v.).

End of 'Illustrated Present Books'

1854 The Photographic Primer for use of Beginners in the Collodion Process. By Joseph Cundall.

Sampson Low, 1s.

1854 Twenty Views in Gloucestershire. Photographed by Joseph Cundall.

The Photographic Institution, 168 New Bond Street. 375 × 280 mm.

'These Photographic Views were taken at the request of an eminent Engineer, to be laid, in Evidence, before a Committee of the House of Commons. They were produced by the Collodion process, in the third week of March, and were all developed in a Post-chaise.'

1855 Photographic Views of the Progress of the Crystal Palace, Sydenham, by Philip H. Delamotte.

Published for the Directors of the Crystal Palace Company, at the Photographic Institution, 168 New Bond Street. 499 × 363 mm. 8 pp. + 100 photographs pasted down on recto pages.

1855 The Calotype Process. A Handbook to Photography on Paper. By Thomas Sutton, B.A., Caius Coll. Cambridge.

Joseph Cundall at the Photographic Institution, 168 New Bond Street. Sampson Low & Son, 47 Ludgate Hill. 183 × 115 mm. viii + 92 pp + 12 pp advts. Printed by G. Barclay.

National Library of Scotland

1855 The Vicar of Wakefield by Oliver Goldsmith. Illustrated by George Thomas. Ornamental decorations by T. Macquoid.

Published for Joseph Cundall by Sampson Low & Son, 47 Ludgate Hill. 200 × 133 mm. viii + 220 pp + 8 pp advts. Printed by R. Clay.

1855 †Examples of Ornament. Selected chiefly from Works of Art in the British Museum, The Museum of Economic Geology, The Museum of Ornamental Art in Marlborough House, and the New Crystal Palace. Drawn from original sources, by Francis Bedford, Thomas Scott, Thomas Macquoid, and Henry O'Neill, and edited by Joseph Cundall.

Bell & Daldy, 186 Fleet Street. 333 × 246 mm. Letterpress prelims + 24 leaves of captions printed one side only by G. Barclay, Castle St, Leicester Square. Title-page on wood in two colours by Leighton Bros. 6 chromolithographic plates by Day & Son. 18 plates lithographed.

1855 Poor Margaret. A tale. Translated from the German of Johanna Schopenhauer; and illustrated with thirteen original drawings by Agnes Fraser.

London: Published by the Committee of the Patriotic Art Exhibition, in aid of the fund for the relief of the widows and orphans of The British Officers who fell in the war with Russia. 208 × 272 mm. iv + 20 pp + 13 mounted photographs of drawings, one to a page.

Note: 'The drawings in this book were photographed gratuitously by Messrs Cundall and Howlett, of the Photographic Institution, New Bond Street, and the impressions were supplied without profit. The paper and the printing of the text were charged at the cost price by Mr George Barclay, 28 Castle Street, Leicester Square. ₓ★ₓ orders for this book will be received at the Photographic Institution, 168 New Bond Street.'

Courtesy Justin G. Schiller Ltd.

c. **1856** BOOKS FOR YOUNG READERS

Price Half-a-crown each.
Addey & Co., Henrietta Street, Covent Garden.

★1. **Amusing Tales for Young People** by Mrs Harriet Myrtle. With 21 pictures.

★2. **The Donkey's Shadow and other stories** with 60 pictures.

3. **The Broken Pitcher and other stories** with 35 pictures (by Harrison Weir) c. 152 pp. Two contributions are initialled 'J.C.' (Item 508 in Justin Schiller Catalogue 29.)

★4. **The Little Lychetts.** A piece of autobiography by author of 'Head of the Family', etc. With 22 pictures.

5. **Historical Tales**
by M. J. With 20 pictures by George Thomas (woodcuts). 166 pp.

6. **The Great Wonders of the World** by Arthur C. Wigan. With 32 pictures by Frederick Skill (woodcuts). 134 pp.

★7. **Visits to the Zoological Gardens** by Frederica Graham.

★8. **The Richmonds' Tour in Europe** by Alfred Elwes. With 28 pictures.

Nos. 5 and 6 both measured 195 × 133 mm, were both printed by William Stevens, 37 Bell Yard, Temple Bar, and were both issued in paper on boards printed with richly decorative designs chromolithographed in 8 colours on front and back, unsigned. The designs combine historical ornament and illustration, and both titles are completely different. The inside pages are commonplace, but Cundall's contributions to No. 3, the cover designs, and Cundall's publication of Arthur Wigan's *The Great Wonders of Art* in 1850 (q.v.) make it probable that Cundall had a connection with the series.

End of Books for Young Readers.

1856 †**Sabbath Bells** *Chimed by the Poets.* Illustrated by Birket Foster.
Bell & Daldy, 186 Fleet Street. 225 × 158 mm. 112 pp. Printed by C. Whittingham. Wood blocks printed in colours by Edmund Evans, with initial letters coloured by hand.
Cundall's monogram (hand coloured) is on back of the title-page. 'Selected by the Editor of "The Poetry of the Year", "The Poets of the Woods", etc.' (at foot of contents page).
The first book with illustrations engraved and printed in full colour by Edmund Evans.
1861: Second edn, published by Bell & Daldy.
1861: A 'new edn', published by Joseph Cundall.

1862: A 'new edn', published by Bickers & Bush.
n.d.: A 'new edn' published by Ward, Lock, & Tyler.
Sabbath Bells appears in a list of new and forthcoming books published by Sampson Low, at the back of *River Gardens* by H. N. Humphreys, Sampson Low, 1857, but I have seen no copy bearing their imprint.

1856 A Photographic Tour among the Abbeys of Yorkshire by P. H. Delamotte & J. Cundall, with descriptive notices by John Richard Walbran FSA and Wm. Jones FSA.
Bell & Daldy. Contains 23 photographs. *Library of The Royal Photographic Society, London.*

1857 The Barefooted Maiden. From the German of Berthold Auerbach. Illustrated by Edward H. Wehnert.
Sampson Low. 162 × 102 mm. viii + 212 + plates + 4 pp adverts. Printed by C. Whittingham, Tooks Court.
The adverts are entitled 'Series of Choice Illustrated Books (Mr Cundall's editions)'.

1857 †**Rhymes and Roundelayes in Praise of a Country Life.** Illustrated by J. Absolon, B. Foster, H. Weir and others, with ornamental headings by T. Macquoid and Noel Humphreys. J. C. monogram on verso of title-page on second edition (same printing with cancel title) of 1857.
D. Bogue, 86 Fleet Street. 220 × 160 mm. viii + 192 pp. Printed by R. Clay.
1875: Reissued by George Routledge.

1858 Poems & Songs by Robert Burns. Illustrated by B. Foster, J. Archer, C. W. Cope, H. Weir, J. C. Horsley & others. Decorations by W. H. Rogers. J. C.'s monogram on verso of title-page, with note facing: 'This selection from the Poetical Works of ROBERT BURNS includes such of his popular POEMS as may with propriety be given in a volume intended for the Drawing-Room; and nearly all the SONGS which are usually published.'
Bell & Daldy, 186 Fleet Street; Edinburgh: J. Menzies. 220 × 162 mm. xviii + 272 pp.
1860. Republished by W. Kent & Co., without J. C.'s monogram but with the note as above, with his initials added.

1858 Red Field, or a Visit to the Country.
A Story for Children. Illustrated by John
Absolon.

Bell & Daldy. 172× 127 mm. Printed by R.
Clay.

Copy inscribed 'Rose Cundall with Mama's
Love, Christmas 1858' found among Cundall's
own books.

**1858 The Poetical Works of Edgar Allan
Poe.** Illustrated by F. R. Pickersgill, John
Tenniel, Birket Foster, Felix Darley, Jasper
Cropsey, P. Duggan, Percival Skelton, A. M.
Madot. (on p xiv) 'Under the superintendence
of Joseph Cundall'.

Sampson Low, Son & Co., 47 Ludgate Hill.
220× 146 mm. xxxii, 248 pp. Printed by R.
Clay.

n.d. (c. 1865): Republished by Ward, Lock
& Tyler, still bearing Cundall's name.

1858 The Story of the Pilgrim's Progress
told for young people. Preface signed J. C.,
St John's Wood, Oct. 1857. With 16 large
illustrations by Edward Wehnert, on wood.

Sampson Low, Son & Co. 175× 124 mm.
160 pp. Printed by G. Barclay, Castle St,
Leicester Square.

Very good typography similar to Home
Treasury Series.

1860. Republished as *The Children's Pilgrim's
Progress* by Bell & Daldy, 186 Fleet Street.

The following titles appeared apparently
uniform with *The Children's Pilgrim's Progress*,
and therefore probably edited and/or planned
by Cundall:

*The Children's Picture-Book of the Life of
Joseph.* Bell & Daldy, 1861.
The Children's Picture-Book of Quadrupeds.
Sampson Low, 1860.
A Picture-Book of Merry Tales. Bosworth &
Harrison, n.d.
*The Children's Picture-Book of Useful
Knowledge.* Bell & Daldy, 1862.
*The Children's Picture-Book of the Sagacity of
Animals.* Sampson Low, 1862.

**1859 The Home Treasury of Old Story
Books.** Illustrated with Fifty Engravings by
Eminent Artists.

[on verso of title] *This volume includes all the
Fairy Tales which were published in the 'Home
Treasury', edited by Felix Summerly, and all the
'Old Story Books of England', edited by Ambrose
Merton.*

Sampson Low, Son & Co., 47 Ludgate Hill.
168× 120 mm. viii+ 288 pp. Printed by G.
Barclay, Castle St, Leicester Square. 51 full-
page engravings on wood, mostly as used in
the Home Treasury and Gammer Gurton
booklets.

Contents:	Artist
Jack the Giant Killer: Part I.	H. J. Townsend.
Jack the Giant Killer: Part II.	H. J. Townsend.
Little Red Riding Hood.	T. Webster.
Jack and the Beanstalk.	C. W. Cope.
Beauty and the Beast.	J. C. Horsley.
Dick Whittington and his Cat.	T. Dalziel.
The Sleeping Beauty in the Wood.	J. Absolon.
Puss in Boots.	E. H. Wehnert.
Cinderella, or the Little Glass Slipper.	J. Absolon.
Peter the Goatherd.	E. H. Wehnert.
Guy Earl of Warwick.	F. Tayler.
Sir Bevis of Southampton.	F. Tayler.
Tom Hickathrift the Conqueror.	F. Tayler.
Friar Bacon.	J. Franklin.
The King and the Cobbler.	J. Absolon.
Patient Grissell.	J. Franklin.
The Princess Rosetta.	J. Absolon.
Robin Goodfellow.	J. Franklin.
The Sisters.	R. Redgrave.
Golden Locks.	J. C. Horsley.
Grumble and Cheery.	C. W. Cope.
The Eagle's Verdict.	F. Tayler.

Ballads.

Chevy Chase.	F. Tayler.
Robin Hood and Little John	F. Tayler.
The Blind Beggar's Daughter of Bethnal Green.	J. Absolon.
The Babes in the Wood.	J. Franklin.
The Death of Fair Rosamund.	J. Absolon.

Of these I have never seen *Puss in Boots* or
Peter the Goatherd in any original Home
Treasury publication or contemporary list.

(See *A Book of Stories from the Home
Treasury*, n.d. c. 1847, p. 65.)

n.d. (late 1850's) A Treasury of Pleasure Books for Young People with elegant illustrations printed in oil colours. 60 illustrations printed in colours from wood.

New York: Hurd & Houghton, Publishers. n.d. (inscribed 1873). 182× 130 mm. Pages not numbered. Contains 17 items, of which all except 4 appeared in the first edition of *A Treasury of Pleasure Books*, 1850. Most are illustrated with the original illustrations, printed in colours from wood (rather crudely).

The first item, 'Large Letters for the Little Ones', consists of 26 full-page pictorial letters printed in colours, which look English, but which I have never seen elsewhere.

1859 The Poets of the West. A selection of favourite American Poems. Illustrated by F. O. C. Darley, Jasper Cropsey, J. H. Hill, B. Foster, etc.

Sampson Low, Son & Co., 47 Ludgate Hill. 216× 150 mm. Printed by R. Clay.

No mention of Cundall anywhere, but book is very much in his style and almost uniform with the other anthologies published by Sampson Low and others which bear Cundall's name.

1859 The Poetical Works of Thomas Gray. Illustrated by B. Foster, with ornamental head- and tail-pieces drawn by W. Harry Rogers, engraved on wood by Edmund Evans.

Bickers & Son. 146× 102 mm. Printed by R. Clay.

n.d.: A later edition published by Sampson Low (also printed by R. Clay).

Stated by H. M. Cundall to have been produced by his father.

1859 Favourite English Poems of the two last centuries, unabridged with upwards of 200 engravings on wood by various artists. Foreword dated St John's Wood, September 1858. J. C.'s monogram on verso of title-page.

Sampson Low, Son & Co., 47 Ludgate Hill. 222× 155 mm. xx+ 380 pp. Printed by R. Clay.

See *Favourite English Poems*, 2 vols, 1863 (p. 86.)

1859 The Children's Picture Gallery. Printed by Leighton Brothers, MDCCCLIX. Introduction signed J. C. Basically same book as *Cabinet Pictures by Modern Painters*, 1862 (p.86).

1859 ★The Children's Picture Book of English History.
Bell & Daldy.

1859 Child's Play by E. V. B.
Sampson Low, Son & Co., 47 Ludgate Hill. 185× 133 mm. 16 colour plates printed from wood. No mention of printer and no mention of Cundall.

The same illustrations were reprinted in an edition published by Sampson Low in 1865 with the colour plates printed inside border tints signed 'W. Dickes'.

The reasons for supposing that Cundall had anything to do with these books are that the first (?) edition of *Child's Play* was published by Addey & Co. in 1853; that Cundall worked on many books for Sampson Low; that a hand-coloured plate from *Child's Play* (1853 edition) was mounted in a Cundall scrap-book; and that *Child's Play* and *A New Child's Play* (1879) are included in a list of Cundall's publications in his family papers.

1859 The Poems of Oliver Goldsmith.
Edited by R. A. Willmott, with illustrations by Birket Foster and decorations by H. Noel Humphreys.

George Routledge & Co. 222× 158 mm. 40 colour plates printed from wood in about 6 colours, with some hand colouring, by Edmund Evans.

1860: 2nd edition, with 52 colour plates.

1877: A New Edition, same colour plates, also printed by Evans and published by Routledge.

Cundall's son H. M. Cundall, in his *Birket Foster*, 1906, crediting his father with the publication of *The Poetical Works of Oliver Goldsmith*, 1851, says that the illustrations were 'elaborately printed in colours'. Since in that book the illustrations were only monochrome, one wonders if this fine colour-printed book was also Cundall's. There is no positive evidence. Such an elaborately-conceived edition would surely have contained some reference to Cundall if he had planned it. The typography has no 'style', but neither does *Favourite Modern Ballads*, 1860 (also printed by Evans), which Cundall did edit. Evans, in his *Reminiscences*, says he did this book for Routledge, and does not mention Cundall in connection with it.

1860 †A Book of Favourite Modern Ballads. Foreword signed by J. C. as Editor. Illustrated by Birket Foster, J. C. Horsley, Samuel Palmer, William Harvey and other artists with wood-engravings in black and a tint, and borders and decorations by Albert H. Warren on every page printed in gold.

W. Kent & Co. (late D. Bogue) 86 Fleet Street. 238 × 178 mm. Printed by Edmund Evans.

Illustrated on pp 20, 21.

†c. 1865. Republished by Ward, Lock & Tyler, identical except that all the illustrations were printed from wood by E. Evans in 6 or 8 colours, including the magnificent double-spread title-page. This colour-printed version was also issued in two halves, as *Choice Pictures and Choice Poems* and *The Illustrated Poetical Gift Book*, both undated.

1860 The Most Excellent Historie of the Merchant of Venice by William Shakespeare. Illustrated by Birket Foster, G. H. Thomas, H. Brandling, with decorations by Harry Rogers. J. C.'s monogram on verso of title-page.

Sampson Low, & Co., 47 Ludgate Hill. 222 × 159 mm. viii + 96 pp. Printed by R. Clay.

Editorial note: 'In offering an illustrated Edition of one of Shakespeare's immortal Plays as a Gift-book for Families, the Editor has considered it to be his duty to omit a few lines which, in the present age, might be thought objectionable'.

Also published in 1860 by D. Appleton & Co., 346 & 348 Broadway, New York.

1860 Songs for The Little Ones at Home. Illustrated with 16 Coloured Pictures by Birket Foster and John Absolon.

Sampson Low, Son & Co., 47 Ludgate Hill. 184 × 133 mm. Printed by G. Barclay. The 16 colour plates were probably printed from wood by Leighton Bros.

1863. The same book with greatly enlarged text was published by Joseph Cundall, 168 New Bond Street. Printed by R. Clay.

1861 Poets' Wit and Humour selected by W. H. Wills. Illustrated with 100 engravings from drawings by Charles Bennett and George H. Thomas.

Bell & Daldy, 186 Fleet Street. 222 × 159 mm. Printed by Whittingham & Wilkins, Chiswick Press. No mention of J. C.

n.d. Poets' Wit and Humour selected by W. H. Wills. Illustrated with 100 engravings from drawings by Charles Bennett and George H. Thomas.

Joseph Cundall, 168 New Bond Street. 222 × 159 mm. viii + 284 pp. Printed by Petter & Galpin, Belle Sauvage Printing Works, Ludgate Hill, E.C.

It is probable that the earliest edition was printed by the better printers (Chiswick Press) and that Cundall took the risk of the second edition and went to a cheaper printer – but it may have been the other way about.

A 'New Edition, just ready' was announced in 1864 advertisements by Ward & Lock.

1861 A Poetry Book for Schools (new edn). 20 large wood-engravings by various artists. Original edition 1855? (date of preface). Recorded as J. C.'s in his family papers, and among his own collection of books.

Bell & Daldy. 175 × 115 mm. Printed by R. Clay.

1861 The Poetry of Nature. Selected and illustrated by Harrison Weir.

Sampson Low, Son & Co., 47 Ludgate Hill. 220 × 155 mm. 112 pp. Printed by Edmund Evans.

Stated to be edited by Joseph Cundall in Jackson's *Treatise on Wood Engraving*, 2nd edn, 1861, p 555.

1861 The Boy's Book of Ballads illustrated with 16 engravings on wood from drawings by John Gilbert.

Bell & Daldy, 186 Fleet Street. 182 × 133 mm. viii + 188 pp. R. Clay, Printer, Bread Street Hill.

No mention of J. C. but copy with his signature is in Cundall family. The text of some of the ballads, e.g. 'Valentine & Orson', is that previously used by J. C., e.g. in *The Playmate*.

1861 Turner's Liber Studiorum. Photographs from the thirty original drawings by J. M. W. Turner R.A. in the South Kensington Museum. Published under the authority of the Department of Science and Art.

Cundall, Downes & Co., 168 New Bond Street. Manchester: Agnew & Sons. 444 × 316 mm.

1862 Cabinet Pictures by Modern Painters.

Joseph Cundall, 168 New Bond Street. 407× 292 mm. Printed by Leighton Bros, and issued in paper covers, with a design on the front by Albert H. Warren, printed in colours from wood by Leighton Bros.

The book consists of large monochrome woodcuts after paintings, which might have been published in the *Illustrated London News* or elsewhere. See *The Children's Picture Gallery*, 1859 (p. 84).

1862 †Shakespeare's Songs and Sonnets. The Foreword is signed by 'J. C.', St John's Wood, November 1861. Illustrated by John Gilbert.

Sampson Low, Son & Co., 47 Ludgate Hill. 348× 248 mm. viii+ 32 pp printed in two colours+ 10 pasted down chromolithographs. Text printed by Edmund Evans, chromolithographs drawn on stone and printed by Vincent Brooks.

The chromolithographs are of superb quality.

This book was also published in smaller format, 215× 146 mm, by Sampson Low, without the colour plates, and the wood engravings and type printed in brown on flat buff tints with highlights in white.

The woodcuts by John Gilbert also appeared in *Shakespere's* [sic] *Songs & Sonnets*, published by Sampson Low in 1863 in the 'Illustrated Present Books' series edited by Cundall, see above, p 80.

1862 The Children's Picture-Book of the Sagacity of Animals illustrated with sixty engravings by Harrison Weir.

Sampson Low, Son & Co., 47 Ludgate Hill. 168× 118 mm. xii+ 276 pp. Printed by Strangeways & Walden, 28 Castle St, Leicester Square. Incl. t.p., 61 hand-coloured woodengravings mostly signed 'J. Greenaway'.

See *The Children's Pilgrim's Progress*, p. 83.

1862 The Poets of the Elizabethan Age illustrated with 30 engravings on wood.

Sampson Low, Son & Co., 47 Ludgate Hill. 220× 160 mm. 84 pp. Printed by R. Clay.

Copy in possession of Cundall family inscribed in Cundall's hand 'To Eliza Cundall with the Editor's love, Christmas 1861'. Advertised as '*Cundall's Elizabethan Poetry*'.

1862 Favourite English Poems of Modern Times illustrated with upwards of two hundred engravings, from drawings by the most eminent artists. J. C.'s monogram on verso of title-page. Foreword dated 'St John's Wood, October 1861' ends with the words 'The Editor hopes shortly to publish a Companion volume of Favourite Poems of the two preceding centuries, uniform in every respect with the present Volume'.

Sampson Low, Son & Co. 220× 160 mm. xii+ 372 pp. Printed by R. Clay.

1863 republished as vol. II of *Favourite English Poems*, q.v.

1863 Favourite English Poems. vol. I. Chaucer to Pope, vol. II Thomson to Tennyson, illustrated with wood engravings by various artists.

Sampson Low, Son & Co. 47 Ludgate Hill. 226× 155 mm. vol. I, xii+ 308 pp. vol. II, xii+ 372 pp. Printed by R. Clay.

No mention of Cundall, but Forewords to each volume are dated 'St John's Wood, 1862' and (vol. II) '1861'.

1863 Early English Poems, Chaucer to Pope. Chiefly unabridged. J. C.'s monogram on verso of title-page.

Sampson Low. 226× 155 mm. xii+ 308 pp. Printed by R. Clay.

This is exactly the same book as vol. I preceding.

1864 Old English Ballads with fifty illustrations by B. Foster, J. Nash, J. Gilbert, H. Hendrick, F. Tayler, G. Thomas, J. Absolon and J. Franklin.

Ward and Lock, 158 Fleet Street. 224× 165 mm. xii+ 272 pp+ 4 pp advts. Printed by Clay.

Stated to have been supervised by J. C. by his son, H. M. Cundall, in *Birket Foster*, 1906. The illustrations are all woodcuts from earlier Cundall editions, including Home Treasury.

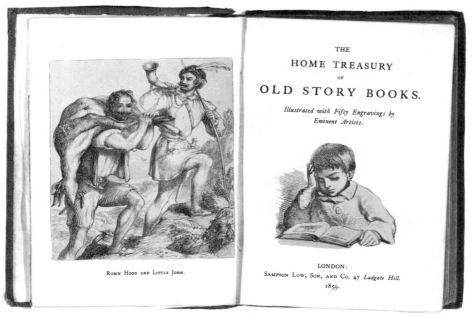

The Home Treasury of Old Story Books, 1859. Title-page opening. Printed by G. Barclay in the Cundall style, using Old Face types. Reduced.

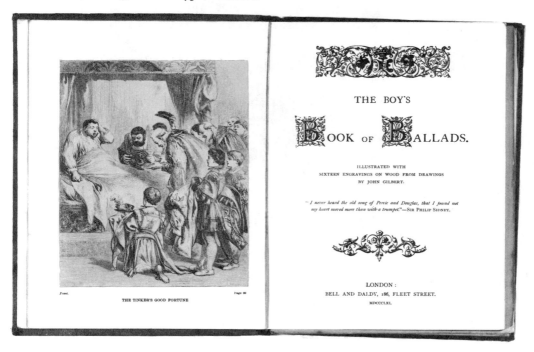

The Boy's Book of Ballads, 1861. Title-page opening with illustration by John Gilbert. Printed by R. Clay, possibly trying to imitate the Chiswick Press decorative style. The decorations, cut on wood, were drawn by Harry Rogers. Reduced.

Joseph Cundall, photographed by Lock & Whitfield, date unknown. Cundall's monogram, probably in his own hand, at right. From a print in the Cundall family.

1864 Portraits of the Members and Associates of the Society of Painters in Watercolours. Portfolio of 36 photographs, with printed title-page. Photographed and published by Cundall, Downes & Co., 168 New Bond Street.

1866 Puss in Boots with original illustrations by H. L. Stephens.

Sampson Low, Son & Marston, Melton House, Ludgate Hill. 184×140 mm. Printed in oil colours by J. Bien. 6 plates in rather poor chromolithography.

There is no proof of any connection with Cundall, but it looks like a Cundall book, with the title-page set in Old Face, and Cundall was working for Sampson Low at this time.

1866 The Great Works of Raphael. 20 photographs by Cundall & Fleming, pasted down.

Bell & Daldy. 186 Fleet Street and Cundall & Fleming, 168 New Bond Street. 245×178 mm. Text printed by Whittingham & Wilkins at the Chiswick Press.

1870. 4th edn. Edited by Joseph Cundall. 20 photographs pasted down 'photographed by the Permanent Printing Company, 9A Hereford Square, S.W.'

Bell & Daldy, York Street, Covent Garden. Cambridge, Deighton, Bell & Co. 274×205 mm. x+76 pp+20 plates pasted down on one side of paper. Text printed by Whittingham & Wilkins.

1866 The Royal House of Tudor by George Wallis. Illustrated with photographs by C. Thurston Thompson, pasted down.

Cundall & Fleming, 168 New Bond St.; Sampson Low, Son & Marston, Ludgate Hill. Printed by Whittingham & Wilkins.

1866 In the Fir-Wood by E. V. B. Photographs of drawings by E. V. B., pasted down.

Cundall & Fleming, 168 New Bond Street. 181×136 mm. 32 pp+8 plates. Printed by Strangeways & Walden, Castle St, Leicester Square.

1866 Photographs of Twelve Drawings by Birket Foster.

Cundall & Fleming, 168 New Bond Street. 280×362 mm (landscape). The title-page and

contents page were printed at the Chiswick Press, but there is no other text. The photographs were pasted down.

1866 Flemish Relics by F. G. Stephens. Illustrated with 15 photographs by Cundall & Fleming, pasted down. Dedicated by the author to W. M. Rossetti.

Alfred W. Bennett, 5 Bishopsgate Without. 241×158 mm.

The typography of this book is not in Cundall's style.

c. 1866 *Normandy by F. G. Stephens. Illustrated with twenty-five photographs by Cundall & Downes.

Alfred W. Bennett, 5 Bishopsgate Street Without. Demy 8 in., elegant binding, 21s. (Advt. at back of *Flemish Relics*.)

1867 Memorials of William Mulready. Collected by F. G. Stephens (author of *Flemish Relics*, etc.). Illustrated with 14 photographs of his most celebrated paintings, professionally coloured. Preface states that: 'the pictures . . . the most important of which have been fortunately reproduced by Messrs. Cundall and Fleming' and: 'Among those who have so kindly assisted me, may I venture to name Messrs H. Cole, C. B.; . . .'

Bell & Daldy, 186 Fleet Street. Cambridge: Deighton, Bell & Co. 276×210 mm. x+134 pp+plates (hand-coloured). Printed by Chiswick Press: Whittingham & Wilkins, Tooks Court, Chancery Lane.

1867 †Two Centuries of Song edited by Walter Thornbury. Illustrated by original pictures of eminent artists drawn and engraved especially for this work, with coloured borders, designed by Henry Shaw, F.S.A. etc., etc. Cundall is mentioned as art director of the book on p vi.

Sampson Low, Son & Marston, Milton House, Ludgate Hill. 228×165 mm. xii+308 pp +20 plates printed one side of sheet only, of which 19 have a flat tint. Printed by R. Clay. The borders by Shaw are printed in red or brown throughout.

1867 The Life and Genius of Rembrandt: The Most Celebrated of Rembrandt's Etchings edited by Joseph Cundall. Thirty photographs (by Cundall & Fleming) taken from the collections in the British Museum, and in the possession of Mr Seymour Haden, pasted down.

Bell & Daldy, 186 Fleet Street; Cambridge, Deighton, Bell & Co. 274×210 mm. Text printed by Whittingham & Wilkins at the Chiswick Press.

1868: 2nd edn.

1868 Rambles in the Rhine Provinces by John P. Seddon. Illustrated with chromo-lithographs, photographs and wood engravings.

14 photographs by Cundall & Fleming pasted down.

John Murray, Albemarle Street. 274×206 mm. Text printed by Whittingham & Wilkins at the Chiswick Press.

The whole book, with an elaborate and highly decorative cover-brass, looks as if it was produced by Cundall.

1868 The Great Works of Sir David Wilkie. 16 photographs by Cundall & Fleming, pasted down.

Bell & Daldy, York Street, Covent Garden; Deighton, Bell & Co., Cambridge. 276×210 mm. The text printed by Whittingham & Wilkins at the Chiswick Press.

Joseph Cundall's signature, c. 1855.

Joseph Cundall

IV. LAST YEARS 1868-95

1869 The Great Works of Raphael: Second Series. Edited by Joseph Cundall. Illustrated with 26 photographs [by Cundall & Fleming?], pasted down.

Bell & Daldy, York Street, Covent Garden; Cambridge, Deighton, Bell & Co. 245×178 mm. Text printed by Whittingham & Wilkins.

1870 An Artist's Alphabet by Godfrey Sykes.

Bell & Daldy, York Street, Covent Garden. 184×127 mm. Letterpress by Chiswick Press. 'Printed in autotype, by Messrs Cundall & Fleming (under licence of the Autotype Printing and Publishing Company, Limited).'

1870 The Wood-Nymph by Hans Christian Andersen. Translated from the Danish by A. M. and Augusta Plesner.

Sampson Low, Son & Marston. 168×124 mm. 60 unnumbered text pages+3 colour plates printed in four colours from wood, unsigned. Printed by Whittingham & Wilkins, Tooks Court, Chancery Lane.

No mention of J. C. but probably produced under his care.

1870 Marvels of Glass-Making in All Ages by A. Sauzay.

Sampson Low, Son & Marston. 196×134 mm. xx+272 pp+pasted down plates. Text printed by William Clowes & Sons.

'Eight of the illustrations in this volume are Reproductions by the Autotype process, printed by Messrs Cundall and Fleming, under licence of the Autotype Printing & Publishing Company (Limited).'

1870 The History of the Life of Albrecht Dürer by Mrs Charles Heaton.

Macmillan & Co. 254×165 mm. xvi+340 pp+10 plates. Text printed by R. Clay, Sons, & Taylor.

'Ten of the illustrations in this volume are reproductions by the Autotype (carbon) process, and are printed in permanent tints by Messrs Cundall and Fleming, under licence from the Autotype Company, Limited.'

1871 The Raffaelle Gallery, etc. Twenty Autotype reproductions.

Bell & Daldy, York St, Covent Garden; Cundall & Fleming. 168 New Bond Street. 368×266 mm. Printed by Whittingham & Wilkins.

1872 The Owl and the Pussy Cat and other Nonsense Songs. Illustrated by Lord Ralph Kerr.

Cundall & Co., 168 New Bond Street. 266×368 mm landscape. 11 photographs of drawings pasted down on card.

Probably published at the artist's expense.

1873 My Lady's Cabinet decorated with drawings and miniatures. Preface signed 'J. C., Bournemouth, November 1872'.

Sampson Low, Marston, Low & Searle. 305×242 mm. Autotypes, sometimes 3 to a page, of paintings and engravings pasted down inside drawn 'frames'. Unnumbered pages. Letterpress by Chiswick Press.

n.d. Works of Art in Pottery, Glass and Metal, in the collection of John Henderson, MA, FSA, photographed and printed by Cundall & Fleming for private use.

495×381 mm. 20 plates. Text printed by Whittingham & Wilkins.

1874 Elementary History of Art by N. D'Anvers (pseud. of Nancy R. E. Bell). Illustrated with 120 woodcuts. 668 pp. Asher & Co., 13 Bedford Street, Covent Garden. 1882 2nd edition published by Low, Marston, Searle & Rivington. The copyright, plant, etc of this work were owned by Cundall and sold by his widow to Sampson, Low, Marston & Co in a Deed dated 24 June 1895.

1875 The Bayeux Tapestry reproduced in Autotype plates. With Historic Notes by Frank Rede Fowke. The Photographs were taken from the Tapestry at Bayeux under the direction of Mr J. Cundall; and have been printed in Autotype by Messrs Spencer, Sawyer & Bird.'

The Arundel Society, 24 Old Bond Street. 284×220 mm. viii+192 pp+79 autotype plates printed one side of paper. Text printed by Whittingham & Wilkins, Chiswick Press.

1877 Painters of all Schools. By Viardot and others.

Sampson Low.

1879-91 Illustrated Biographies of the Great Artists. 39 volumes edited by Joseph Cundall, published by Sampson Low, and advertised in 1880 at 3s 6d per volume.

*Illustrated Handbooks of Art History. 5 volumes edited by Joseph Cundall.

1879 Hans Holbein ed. by Joseph Cundall. In the Illustrated biographies of the Great Artists series. Text from *Holbein und seine Zeit* by Dr Alfred Woltmann.

Sampson Low, Marston, Searle & Rivington, Ltd. 188×130 mm. xii+116 pp+16 plates. Printed by R. Clay, illustrated with many wood-engravings, and Chiswick Press tail-pieces. The plates are monochrome wood-engravings reproduced by lithography.

1892: reprinted.

1881 On Bookbindings Ancient and Modern edited by Joseph Cundall. Dedicated to Sir Henry Cole 'my earliest and kindest instructor on all questions of art, with sentiments of deep gratitude' by J. C.

George Bell, York Street, Covent Garden. 254×188 mm. Printed by R. Clay.

1883 The Renaissance of Art in Italy by Leader Scott (Lucy E. Baxter). Reference to Mr Cundall the art editor in preface.

Sampson Low. 282×216 mm. xxii+384 pp 4to.

1885 *Don Quixote edited by J. Cundall.

1886 Annals of the Life and Work of William Shakespeare. Preface signed J. C.

Sampson Low, Marston & Co. 8vo. 146 pp. Printed by R. Clay.

1895 A Brief History of Wood-Engraving from its invention, by Joseph Cundall.

Sampson Low, Marston & Co., Fetter Lane, Fleet Street E.C. 184×124 mm. x+132 pp. Printed by Spottiswoode & Co.

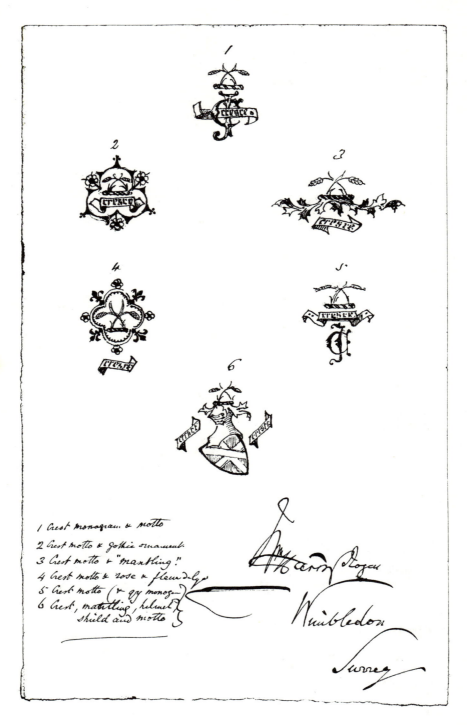

Sketches for a Cundall monogram and crest by W. Harry Rogers, date unknown. From a Cundall family scrapbook.

GENERAL INDEX

Bold figures indicate illustrations

Absolon, John 22, 30, **31**, 42, 49, 50, 52, 54, 55, 57, 58, 61, 62, 65, 66, 69, 71, 72, 75, 77, 82, 83, 85, 86
Addey, H. M., & Co. 22, 26, 72, 75, 76, 77, 78, 79, 81
  bankrupt 75
Agnew & Sons 86
Albert Edward, Prince of Wales 14, 57
Andersen, Hans 18, 61, 66, 90
Angelo, Michael 49, 50
Appel's Anastatic Press 76
Appleton, D., & Co., New York 85
Archer, J. 82
Arundel Society, The 91
Asher & Co. 91
Auerbach, B. 61, 82
Autotype 90, 91

Baldwin & Cradock 57
Bahr, Dr J. 77
Barclay, G. 34, 36, 62, 65, 67, 69, 70, 72, 75, 78, 79, 81, 83, 85, **87**
Barclay, H. 79
Bartlett, J., Cambridge, Mass. 71
Baxter, George 12
Baxter, Lucy E. 91
Bayeux Tapestry 44, 44n, 91
Bechstein, L. 79
Bedford, Francis 42, 81
Bell, George 4, 14n, **15**, 22, 34-35, 57, 72, 78, 79, 91
Bell & Daldy 34, 36, 42, **43**, 81-84, 85, 87, 89, 90, 91
Bell, Nancy R. E. 91
Bennett, Alfred W. 89
Bennett, Charles 85
Berghem 48, 56
Bickers 34
Bickers & Bush 82
Bickers & Son 84
Bien, J. 89
Blackford, Mrs 56, 57
Blashfield, Mr 48
Bogue, David 1n, 26, 30, 36, 40, 42, 62, 66, 67, **68**, 69, 72, 75, 78, 79, 82
  became W. Kent & Co. 85
Bohn, H. G. 22, 47, 57, 58, 67
Bolton, T. 79
Boner, C. 61
Bosworth, Thomas 34, 42, 78, 79
Bosworth & Harrison 83
Boyle, E. V. 76, 78, 79, 89
Brandard, John **10-11**, 16, 58, 67
Brandling, H. 85
British Museum 14, 16, 16n, 22, 33n, 45, 57, 81, 90
Brockhaus, F. A., Leipzig 66
Brooks, Vincent 38, 44, 78, 86
Browne, H. K. (Phiz) 72, 75, 76, 79
Brunel, I. K. 33
Burchett, R. **24**, **25**, 62, 66
Burns, James 47

Cadogan, Lady Augusta 69
Calotype process 78, 81
Carroll, Lewis 33, 33n
Chapman & Hall 18, 26, 48, 52
  Home Treasury bought 18, 48
Chatelain, Mme de 70, 71, 77
Chiswick Press (see Whittingham, Charles, and Whittingham & Wilkins)
Clarke, B. 47, 48
Clarke, Rev. C. 65
Clay, Richard **40**, **41**, 42, 44, 62, 79, 80, 81, 82, 83, 84, 85, 86, 87, 89, 90, 91
Clowes, William 44, 90
Cole, Sir Henry, (1808 82) (pseudonym Felix Summerly) 1, 1n, 4, 12, 14, **15**, 16, 30, 44, 45, 48, 49, 57, 89, 91
  Felix Summerly 4, 48, 49, 52, 65, 83
  Tea Service 14
  Summerly's Art Manufactures 14
  Mrs Felix Summerly 50n, 52
Collins & Co. (see also Gregory Collins & Reynolds, Collins & Reynolds and Reynolds & Collins) 67
Collins & Reynolds (see also Gregory Collins & Reynolds, Reynolds & Collins, and Collins & Co.) **24**, 62
Collodion Process 78, 81
Cope, C. W. 18, 49, 50, 55, 82, 83
Corbould, H. 48, 50, 55

Coutts Bank 1n
Coutts, Miss Burdett 44
Cox, David 78
Cox, J. & H. 47
Creswick, T. 78
Cropsey, Jasper 83, 84
Crosby, Lockwood & Co. 71
Crossthwaite, Miss 1n
Cruikshank, George 1, 1n, 61, 65
Crystal Palace 42
  Coy 81
  New 81
Cundall & Addey 26, **27**, 67, 70, 71, 72, 75
  ad. 71, 72
Cundall & Co. 91
Cundall family vi, 1, 2-3, **11**, 16, 22, 22n, 30, **32**, 42, 42n, 44, 45, 45n, 48, 62, 69, 71, 77, 79, 83-86, 91
  Sarah Ranson **32**, 45
  Emily Thompson 45
  Children 2-3
    Sarah Maria (Maja) 16, 30, 45, 69, 71, 75
    Herbert 22, 22n, 30, 69, 71, 77, 79, 80, 84, 86
    Joseph Henry 16, 30
    Rose 30, 83
    Eliza 86
    Edmund 22
    Edward George 42
    Frank 42, 44
  Will 45
Cundall & Fleming 34, **43**, 45, 89, 90, 91
Cundall, Downes & Co. **32**, 33, 86, 89
Cundall, Howlett & Co. 33, 34, **35**, 81
Cundall, Howlett & Downes 33
Cundall, Joseph (1818-1895) (pseudonym Stephen Percy)
  advertisements 53-56, 89
  bankruptcy 22, 26, 75
  description 16, 45
  1862 Exhibition 42, 44
  family bible 1, 16n
  family scrapbook 62, **63-64**, 84, **92**
  family tree 2-3
  marriages 16, **32**, 45
  monograms iii, **24**, **25**, 34, 36, 38, **39**, 62, **71**, **74**, 75, 80, 82, 84-86, **88**, **92**
  partnerships **32**, 33, 75
    Addey 22, 26, 75
  Percy, Stephen 1, 47, 56
  photography **32**, 33-34, **35**, 78, 81, 82, 89-91
  portraits viii, 33-34, **46**, **88**
  signature **90**
  typefaces used 16, 18, 41, 89
  V & A Museum 44
  Will 45

Daguerrotype 70
Dalziel Brothers 30, 38, 70
  E. 62
  G. & E. 70
  T. 83
Dana, C. A. 66
Dangerfield, F. 18, **25**, 61, 66
Darley, Felix 70, 83, 84
Day & Son 44, 67, 71, 78, 81
Day & Haghe 57
Deighton, Bell & Co. 89, 90
Deighton, J. & J. J. **15**, 57
De la Motte, Philip H. 30, 33, 42, 72, 78, 81, 82
Dickes, W. 38, 67, 80, 84
Dobson, Austin 14n
Dorset, Mrs 69, 70
Dossetter, E. 44
Doyle, Richard 12, 30, 70
Drake, Lawrence 3, 48, 56
Duggan, P. 83
Duncan, Edward 39, 80
Durer, Albert 12, 14, **15**, 48, 49, 55, 56, 57

Edlin, Thomas **13**, 30, 58
Elwes, A. 82
Erxlehen, J. 57
Etching Club, The 16, 16n, 18, 62, 67, 79
Evans, Edmund 12, **20-21**, 34, 36, 38, 42, 44, 80, 82, 84, 85, 86

Everdingen, A. van 48, 49, 50, 56, 57
Eyton, Sir Joseph Walter King 14

Fairstar, Mrs 61
Fawcett, Benjamin 38
Fenton, Roger 78
Fitzroy, Lady C. 62
Fokke, Simon 58
Follen, Mrs 78
Foster, Birket **20-21**, 36, 38, 39, 42, 62, **68**, 71, 78, 79, 80, 82-86
Fowke, F. R. 44, 91
Franklin, J. 50, 52, 83, 86
Fraser, Agnes 81
Frere, A. F. 77
Frith, W. P. 33

Gad & Keningale 62, 67
Georgina, A. 77
Gilbert, John 1, 38, 47, 56, 70, 77, 80, 85, 86, **87**
Gillies, Margaret 61
Gilman, C. 78
Giotto 49, 50
Godwin, James 71, 76, 77
Gordon, Sir A. D. 65
Graham, F. 82
Grant & Griffith 26, 30, 61, 65, 69
Great Eastern (ship) 33
Great Exhibition of 1851 14, 30, 72
Greenaway, J. 86
Gregory, Collins & Reynolds (see also Collins & Reynolds) 12, **23**, 49, 52, 57
Greville, R. N. 65
Griffin, Charles 36, 78
Griffith & Farran 70
Grimm, A. L. 61
Grimm, Ludwig 61
Grimm, M. M. 61
Groombridge 38

Hailes, N. 1, 1n, 4, 26, 47, 69
Hanhart, M. & N. 10, 16, 36, 58, 61, 62, 66, 67, 70, 78
Hannay, J. 76
Harrild, Thomas 58, 62, 66, 70
Harris, J. 26
Harvey, William 62, 65, 85
Hawkshaw, Ann, Lady 75
Hayday, 14, 67
Hazlitt, William 14, 57
Henderson, J. 91
Hendrick, H. 86
Hicks, G. E. 80
Hill, J. H. 84
Holbein, Hans 4, 5, 12, 22, 45, 48, 49, 50, 55
Hooper, J. W. 77
Horsley, J. C. 12, 18, 48, 49, 50, 54-55, 57, 82, 83, 85
Hosemann, Theodor 70
Howlett, Robert 33
Hullmandel 1n
Hulme, F. W. 76
Humphreys, Henry Noel 16, 82, 84
Hurd & Houghton 84
Huttula, A. 80

Imperial Printing Office in Vienna 44

Jardin, Karl du 50
Johnstone, John 62
Jones, Owen 18, 30, 42, 44, 69
Juvenile Library, The 1, 4

K., Mme Emma de 56, 57
Kaulbach, Wilhelm 66
Kelly, W. K. 77
Kent, W. 34, 36, 38, 82, 85
Kerr, Lord Ralph 91
Kronheim 12

Lamb, Charles & Mary 65
Landells, Ebenezer 36
Lee, William 72
Leighton Bros 1n, 12, 33, 38, 42, 44, 71, 72, 81, 84, 85, 86
Leighton, H. 75, 76
Leighton, Jane 1n
Leighton, John (pseudonym Luke Limner) 33, 58, 75, **76**
Levey, Robson & Franklyn (see Robson, Levey & Franklyn)

Limner, Luke (see John Leighton)
Linnell, John 49
Lock & Whitfield 88
Longman, Brown, Green & Co. 18, 18*n*, 22, 52, 67
Low & Son 74, 75
Lydon, A. F. 38

M., F. L. ad 56, 58
McDowall, W. 70, 71 .
Maceroni, E. 71
McGlashan, J. 75
MacLachlan, Stewart & Co. 62, 65
Macmillan & Co. 90
Macquoid, T. 74, 75, 81, 82
Madeley 49
Madot, A. M. 83
Marples, David 22
Meadows, Amy 79
Meadows, Kenny 66, 72, 78
Menzies, J. 75, 82
Merton, Ambrose (see Thoms, W. J.)
Miller, Thomas 72
Minton, Herbert 14, 22, 22*n*
Minton, Messrs 49, 50
Mulready, W. 50, 52
Murray, John 90
Myrtle, Mrs Harriet 26, 57, 58, 70, 72, 79, 81

Nash, J. 86
National Portrait Exhibitions, 1866 & 1867 44
Naylor, S. 18*n*, 58
Neill & Co. 62
Newdegate, F. B. 70
Niebuhr, B 52
Nieritz, G. 77
Norman, G. 61
Norton, Hon. Mrs 61

O'Neill, Henry 81
Owen, Hugh 78

Palmer, Samuel 85
Parker, J. H. 15, 57
Parley, Peter 4
Patriotic Art Exhibition, Committee of the 81
Peacock, Thomas Love 1, 50
Percy, Stephen (see Cundall, Joseph)
Permanent Printing Company 89
Petter, Duff & Co. 72, 75, 77
Petter & Galpin 44, 79, 85
'Phiz' (see Browne, H. K.)
Photographic Club 33, 33*n*
Photographic Institution, The 33, 81
Pickering, William 14, 15, 22, 57
   Bankruptcy 22
Pickerskill, F. 83
Piombo, S. del 49, 50
Plesner, A. M. & A. 90
Plonquet, Hermann 72
Pocci, Count 61, 64
Potter, Paul 50
Prentiss, Mrs E. 79
Prosser's Patent, Mr 49
Prussia, Printer to the Court of 44

Quarles, Rev. T. 47

R., M. J. 77
Raffaelle 48, 49, 55
Randolph, Mrs Charles 26, 71
Redgrave, R. 18, 49, 55, 83
Rembrandt 50
Remnant & Edmonds 16
Reynell & Weight 57, 61
Reynolds & Collins (see Gregory Collins & Reynolds and Collins & Co.) 65
Richards 58
Richter, Lewis 66, 79
Rider, J. & W. 78
Robson, Levey & Franklyn 61, 65, 70, 71, 72, 79
Rogers, W. Harry 34, 71, 82, 84, 85, 87, 92
Rosling, Alfred 78
Routledge, George 34, 38, 72, 75, 82, 84
Royal Academy 12
Royal College of Art 44
Royal Photographic Society of Great Britain (previously Photographic Society) 33, 35
Royal Society of Arts 14

Sampson Low 26, 31, 34, 36, 38, 39, 42, 45, 52, 67, 70, 75, 79, 79, 80-87, 91
Low, Marston, Searle & Rivington 91
Sampson Low, Marston & Co. 91
Sampson Low, Marston Low & Searle 91
Sampson Low, Marston, Searle & Rivington 91
Sampson Low, Son & Marston 89, 90
Sauzay, A. 90
Schneider, H. J. 72
Scott, Thomas 81
Sharpe & Hailes 1
Sharpe, John 1, 4
Shaw, G. 78
Shaw, Henry 12, 22, 42, 89
Simms, Charles 70
Skelton, Percival 83
Skill, Frederick 61, 82
Slader, S. V. 79
Slater, Mrs John 57
Society of Arts 30, 50, 72
Sonderland, J. B. 61
South Kensington Museum, (see Victoria & Albert Museum) 44, 86
Spencer Sawyer & Bird 91
Spottiswoode & Co. 44, 91
Starling, Elizabeth 56, 57
Stephens, F. G. 89
Stephens, H. L. 89
Stereotyping 14
Stevens, William 82
St George's Reading Library 14, 60
Stoddart, Lady 65
Stoop 12, 48, 50, 56
Stothard, T. 13, 56, 58
Strangeways & Walden 86, 89
Strickland, Jane 48
Summerly, Felix (see Cole, Henry)
Summerly, Mrs Felix 50, 52
Sykes, G. 90

Tayler, Frederick 18, 49, 50, 52, 54, 55, 61, 83, 86
Taylor, Edgar 61
Taylor, Jefferys 47, 48
Taylor, R. & J. E. 61, 65
Tegg, William 70
Teniers 12, 48, 50, 56
Tenniel, Sir John 12, 83
Thackeray, William 12
Thomas, George 39, 79, 80, 81, 82, 85, 86
Thompson, Charles Thurston 14, 89
Thompson & Davidson 71
Thompson, L. 79
Thoms, W. J. (pseudonym Ambrose Merton) 12, 48, 50-52, 53, 83
Thornbury, Walter 42, 89
Tilt & Bogue 47
Tilt, Charles 1, 1*n*, 47
Townsend, H. J. 18, 49, 50, 52, 55, 83
Trimmer, Mrs 65
Turner, J. M. W. 86
Twining, E. 67

Victoria & Albert Museum (originally called South Kensington Museum) 44, 44*n*
Vizetelly, Henry 47, 62

Ward & Lock 62, 85, 86
Ward, Lock & Tyler 34, 38, 82, 83, 85
Warde, Mrs W. 69, 73
Warren, Albert H. 20-21, 38, 85, 86
Warren, Arthur 78, 85
Warren, H. 80
Waterford, Marchioness of 67
Webster, Thos. 48, 49, 50, 55, 83
Wehnert, E. H. 50, 71, 72, 76, 77, 80, 80, 82, 83
Weill, A, 65
Weir, Harrison 26, 30, 65, 69, 71, 72, 76, 78-82, 85, 86
Whitehead & Co. 61
Whittingham, Charles the Younger 4, 12, 14, 16, 18, 22, 26, 34, 36, 42, 45, 48, 52, 57, 58, 61, 65-67, 69-72, 76, 78, 82
   work illustrated ii, 6, 7, 13, 15, 17, 23, 28, 29, 53-6, 60, 73
   family 16
Whittingham & Wilkins 43, 85, 89, 90, 91
Wigan, A. C. 70, 82
Wilkes, M. G. 77
Willmott, R. A. 36, 84
Wills, W. H. 85

Wilson, John 61
Wolf, Joseph 36, 37, 78
Wyatt, M. Digby 44

TITLES OF BOOKS

Italic figures indicate an entry in check-list

Aladdin and the Wonderful Lamp and Sinbad the Sailor, 1853 77
L'Allegro, 1849 16, 18, 67
Alphabet of Quadrupeds, An, 1844 6, 48, 50
   ad. 56
Amusing Tales for Young People 81
Ancient History, An 77
Artist's Alphabet, An, 1870 90
Anecdotes of Little Princes ad. 54
Animals from the Sketch Book of Harrison Weir, 1851 72
Annals of the Life and Work of Shakespeare, 1886 45, 91
Apple Dumpling, The 77
Arbell 77
Architectural Pastime, see Box of Terra-Cotta Bricks, 1845 49
Art Album, The, 1861 38
Arts and Arms, see Puck's Reports to Oberon, 1844 50
Aunt Carry's Ballads, 1847 61
Aunt Effie's Rhymes for little children, 1852 75

Babes in the Wood, The, 1849, 1851 67
GAMMER GURTON'S STORY BOOKS, c. 1843 50, 53(ad.), 83
ILLUSTRATED PRESENT BOOKS, 1861 38, 67, 80
Ballad of Chevy-Chase, The, 1844 49, 50, 83
   ads. 55, 56
Ballad of Sir Hornbook or Childe Launcelot's Expedition, The, 1843 48, 50
   ads. 55, 56
Ballads and Faery Tales, c. 1846 ad. 56
Barefooted Maiden, The, 1857 82
Beauty and the Beast, 1843 48, 50, 83
   ads. 55, 56
Bayeux Tapestry, The, 1875 91
BERTIE'S INDESTRUCTIBLE BOOKS 77
Bertie's Farm Yard
Bertie's Foreign Birds
Bertie's Horn Book
Bertie's Wild Beasts
Bertie's Woodside
Bertie's Word Book
Bible Events, 1st, 2nd, 3rd & 4th series, 1843-44 48-50
   ad. 55
Blind Beggar's Daughter of Bethnal Green, The, 1845 50, 52, 83
Book of Animals, The, 1847 26
Book of Favourite Modern Ballads, A, 1860, c. 1865, see Favourite modern Ballads, A, Book of. See also Choice Pictures & Choice Poems and The Illustrated Poetical Gift Book
Book of Ruth, The, 1850 69
Book of Stories from the Home Treasury, A, c. 1847 65
Booke of Christmas Carols, A, [1845], 1846 10-11, 16, 58
BOOKS FOR YOUNG READERS 70, 81-82
Box of Terra-Cotta Bricks, 1845, with Architectural Pastime 49, 50
Boy's Almanac for 1849, The, 1849 67
Boy's Book of Ballads, The, 1861 85, 87
Bridal Gift, A, 1847 22
Brief History of Wood-Engraving, A, 1895 91
Broken Pitcher, and other stories, The, c. 1855 81
Burns, Robert, Poems and Songs by, 1858 82

Cabinet Pictures by Modern Painters, 1862 86
Calotype Process, The, 1855 81
Century of Fables, A, c. 1845 49, 50
Chaplet of Pearls, A, 1851 26, 29, 71
Charm, The, 1853 75
Charm Almanack, The 77
Children's Picture Book of English History, The, 1859 84

94

Children's Picture-Book of the Life of
Joseph, The, 1861 83
Children's Picture-Book of Quadrupeds, The,
1860 83
Children's Picture-Book of the Sagacity of
Animals, The, 1862 42, 83, 86
Children's Picture-Book of Useful Know-
ledge, The, 1862 83
Children's Picture Gallery, The, 1859 84
Children's Pilgrim's Progress, The, 1860
83, 86
Children's Summer, A, 1853 78
Child's Own Alphabet, The, c. 1852 74, 75
Child's Play, 1853 76
1859, 1865 84
Choice Examples of Art Workmanship,
1851 26, 27, 30, 72
Choice Pictures and Choice Poems 38, 85
Church Catechism, The, 1851 71
Cinderella, 1845 49, 50, 83
ads. 55, 56
Cinderella, illus. by M.J.R. 77
Colour Box for Little Painters, c. 1845 49,
50
Comic Alphabet, 1835 1n
Comical Creatures from Wurtemberg, The,
1851 72
Common Wayside Flowers, 1860 38
Cottage Traditions; or, the Peasant's Tale of
Ancestry, 1842 47
Cousin Natalia's Tales, 1841 47
Creed, the Lord's Prayer and the Ten
Commandments, The, 1848 18, 19, 66
Cundall's Elizabethan Poetry see Poets of
the Elizabethan Age, The, 1862
Cundall's Indestructible Books 75, 78

Danish Story-Book, A, 1846 18, 61
Day of Pleasure, A, 1853 72
Deserted Village, The, 1841 16, 18
1855, 79
Diary and Houres of the Ladye Adolie, a
faythfulle Childe, The, 1853 76
Discontented Chickens, The, 1853 78
Diverting History of John Gilpin, The, 1845
58
Doleful Story of the Babes in the Wood, The,
and The Lady Isabella's Tragedy, 1843
50, 83
Don Quixote, 1885 91
Donkey's Shadow and other stories, The,
c. 1856 81

Eagle's Verdict, The, 1844 49, 50n, 83
ad. 56
Early English Poems, Chaucer to Pope, 1863
see Favourite English Poems, vol. 1 38,
86
Edwin Evelyn, A Tale, 1843 48
Ellen Cameron, 1845 57
Encyclopaedia of Ornament, The, 1842 22
Eskdale Herd-Boy, The, c. 1848 65
Elementary History of Art, 1874 91
Euclid, by Oliver Byrne, 1847 52
Eve of St Agnes, The, 1856 80, 80
Events in Sacred History, see also Bible
Events
ad. 55
Examples of Ornament, 1855 42, 81
Excitement, The, 1847 62

Fairy Folk and Wonderful Men, c. 1852 77
Famous History of Friar Bacon, The, c. 1843
50, 52, 83
Famous History of Sir Guy of Warwick,
The, c. 1843 50, 83
Far-Famed Tales, 1852 77
Farmer's Boy, The, 1857 80
Favourite English Poems, 1863 38, 84, 86
Favourite English Poems of Modern Times,
1862 86
Favourite English Poems of the two last
Centuries, 1859 41, 84
FAVOURITE LIBRARY, THE, 65
Favourite Modern Ballads, A Book of, 1860,
c. 1865 20, 21, 34, 36, 38, 84, 85
Feathered Favourites, 1854 36, 78
Flemish Relics, 1866 89
Forty Favourite Fairy Tales 71
Fraser's Magazine, 1846 12
Funny Rhymes and Favourite Tales, c. 1852
77

Gallant History of Sir Bevis of Hampton,
The, c. 1843 50, 52, 83

GAMMER GURTON'S STORY BOOKS. etc,
c. 1843 6, 12, 48, 50, 52, 60, 83
ad. 53
variant titles 52
Gems from the Poets, 1860 38
Gems of English Art, 1869 38
Gems of Nature and Art, c. 1868 38
German Fairy Tales & Popular Stories,
1846 61
Glance at the Exhibition of the Royal
Academy, A, 1850 70
Gleaner, The, 1846 61
Golden Locks, 1844 49, 50, 83
ads. 55, 56
Golden Songs for Silvery Singers, c. 1853 77
Goldsmith, Oliver, The Poems of, 1859 36,
38, 84
Goldsmith, Oliver, The Poetical Works of,
1851 71, 84
Good-Natured Bear, The, 1846 61
ad. 54
Grammar of Ornament, The, 1856 42
Grand Historical Pictures, 1848 66
Gray's Elegy, 1847 16, 62, 79
Gray, Thomas, The Poetical Works of, 1859
84
Great Wonders of Art, The, 1850 70, 82
Great Wonders of the World, The, c. 1856
82
Great Works of Raphael, The, 1866 43, 89
2nd series 90
Great Works of Sir David Wilkie, The,
1868 90
Green Bird, The, c. 1853 77
Grimm's Household Stories, c. 1853 77
Grumble and Cheery, 1844 49, 50, 83,
ads. 55, 56

Hamlet, The, 1859 38, 80
Handbook for the National Gallery, 1843 4
Hans Holbein, 1879, 1892 91
Happy Days of Childhood, 1854 79
HARRY'S LADDER TO LEARNING, 1849-
1850 30, 69
Harry's Country Walk
Harry's Horn-Book
Harry's Nursery Songs
Harry's Nursery Tales
Harry's Picture-Book
Harry's Simple Stories
HAZLITT'S HOLIDAY LIBRARY, 1844
14, 57, 59
ad. 56
Hero, A, c. 1853 77
Heroes of England, The, 1843 48
ad. 56
Heroic Life and Exploits of Siegfried the
Dragon Slayer, The, 1848 17, 25, 66
Heroic Tales of Ancient Greece, 1844 49,
50, 52
ad. 56
Historical Tales, c. 1856 82
History of Antiquities of Foulsham in
Norfolk, The, 1842 47
History of the Life of Albrecht Durer, The,
1870 90
History of the Robins, The, 1848 65
History of Tom Hickathrift the Conqueror,
1845 52, 83
HOLIDAY LIBRARY, THE, 1844 (see
HAZLITT'S HOLIDAY LIBRARY)
Holly Grange, a Tale, 1844 57
ad. 56
Home for the Holidays, 1851 72
Home Pictures, 1851 72
HOME STORY BOOKS, 78
HOME TREASURY, THE, 1843 ii, 4, 5-7,
12, 14, 16, 18, 22, 48-50, 52, 62, 83, 86,
87
ad. 55-56
Home Treasury of Old Story Books, The,
1859 i, 52 (see also 65), 83, 87
Home Treasury Primer, The (see Mother's
Primer, The) 50, 50n, 52
Hours of Day and Spirits of Night, 1847 61

Illuminated Calendar, The, 1845 16
Illustrated Biographies of the Great Artists,
1879-91 44-45, 91
Illustrated Catalogue 1862 International
Exhibition, London, 42, 44
Illustrated Handbooks of Art History, 91
Illustrated London News, 12 April 1856
34n
Illustrated Poetical Gift Book, The 38, 85

'ILLUSTRATED PRESENT BOOKS' 67,
79-80, 86
Illustrations of the Natural Orders of Plants,
1849-55 67
Illustrations to Southey's Roderick, 1850 70
In the Fir-Wood, 1866 89
Indestructible Lesson Book, The, c. 1852 75
Indestructible Reading Book, The, c. 1852
75
Indestructible Spelling Book, The, c. 1852
75
Indestructible Primer, The, c. 1852 75

Jack and the Bean Stalk, 1844 49, 50, 83
ads. 55, 56
Jack and the Giants, The Story of, 1851 30,
70
Jack the Giant Killer, 1845 49, 50, 83
ads. 55, 56

Kindness and Cruelty, c. 1853 77
King of the Swans and other tales, The,
1846 61
ad. 54

Lady Isabella's Tragedy, The, see Doleful
Story of the Babes in the Wood, The
Laughter Book for Little Folk, A, [c. 1851]
70
Legends of Rubezahl, 1845 57, 59
ad. 56
Letters to an Undergraduate of Oxford, 1848
65
Life and Adventures of Robinson Crusoe, The
1845 13, 58
ad. 56
Life and Genius of Rembrandt: The Most
Celebrated of Rembrandt's Etchings, The,
1867 90
Life of Christ, The, 1844 48
ad. 55
Life of Wellington for Boys, A, 1853 77
Little Basket-Maker and other Tales, The,
1845 58
ad. 54
Little Bo-Peep and other Stories, in Rhyme,
1845 58
ad. 56
Little Drummer, The, 1853 77
LITTLE FOLK'S BOOKS 77-78
Little Fortune Teller, The, c. 1853 78
Little Foundling and Other Tales, The, A
Story Book for Summer, 1846 57
Little Lychetts, The, c. 1856 81
LITTLE MARY'S BOOKS, 1847-1850 62
Little Mary's Babes in the Wood
Little Mary's First Book of Original
Poetry 69
Little Mary's First Going to Church
Little Mary's Picture Book of English
History
Little Mary's Primer 8-9, 51, 66-67
Little Mary's Reading Book
Little Mary's Scripture Lessons 68
Little Mary's Second Book of Original
Poetry 68, 69
Little Mary's Spelling Book
Little Mary's Treasury of Elementary
Knowledge, 1865 62
Little Painter's Portfolio, The, c. 1845 49,
50
Little Princes, 1843 57
Little Red Riding Hood, 1843 48, 50, 83
ads. 55, 56
Little Sister, The, 1852 72
Little Susy's Six Birthdays, 1854 79
Lively History of Jack and the Beanstalk,
The, 1844 49, 50, 83
ads. 55, 56

Mad Pranks of Robin Goodfellow, The, 1845
52, 83
Magic Words: A Tale for Christmas Time,
1851 71
Maja's Alphabet, c. 1852 75
Man of Snow and other Tales, The. A Story
Book for Winter, 1848 58
Marvels of Glass-making in All Ages, 1870
90
Memoirs of a London Doll, 1846 61
ad. 54
Memoirs of Bob, the spotted Terrier, 1848
65
Memorials of William Mulready, 1867 89
Merry Tales for Little Folk, 1851 71

Merry Tale of the King and the Cobbler, A,
 1843 *50*, *52*, *83*
Mia and Charlie, 1854 *79*
Miller's Son, The, 1846 *61*
Miss Simmons's Debut, 1848 *65*
Moon's Histories, The, 1848 *18*, *66*
Most Excellent Historie of the Merchant of
 Venice, The, 1860 *85*
Mother Goose, and Simple Simon c. 1852
 75
Mother's Primer, The, 1844 (see Home
 Treasury Primer, The) *49*, *50n*, *52*
Mournful Ditty of the Death of Fair
 Rosamond to which is added Queen
 Eleanor's Confession, A, 1845 *52*, *83*
Mrs Leicester's School, 1848 *65*
My Lady's Cabinet, 1873 *91*
MYRTLE STORY BOOKS, THE, 1845-1848
 57, *58*
 ad. *54*

Naughty Boys and Girls, 1853 *77*
New Child's Play, A, 1879 *84*
New Nursery Songs for All Good Children,
 1852 *78*
New Tales from Faery Land, 1852 *72*
Noble Deeds of Woman, 1845 *57*
 ad. *56*
Normandy, c. 1866 *89*
Nursery Heroes, c. 1852 *77*
Nursery Heroines, c. 1852 *77*
Nursery Sunday Book, The, 1845 *58*
Nut-Cracker and Sugar-Dolly, 1849 *66*

Ocean Child, or Showers and Sunshine, The,
 1857 *26*
Odes & Sonnets, 1859 *38*
Old English Ballads, 1864 *86*
Old Story Books of England, The, 1845
 (see also GAMMER GURTON'S STORY
 BOOKS) *52*, *83*
Old Story-Teller, The, 1854 *79*
On Bookbindings Ancient and Modern, 1881
 16, *45*, *91*
On Ornamental Art applied to ancient and
 modern bookbinding, 1848 *5*, *65*, *66*
Orphan of Waterloo, The, 1844 *57*
 ad. *56*
Owl and the Pussy Cat, The, 1872 *91*

Painters of all Schools, 1877 *91*
Passion of Our Lord Jesus Christ, The, 1844
 15, *57*
Pastoral Poems, 1858 *80*
PATENT INDESTRUCTIBLE PLEASURE
 BOOKS **74**, *75*
Peacock at Home, The, 1849 & 1851 *26*,
 28, *69*, *70*, **73**
Peasant's Tale, The, 1843 *48*
Pentamerone, or The Story of Stories, The,
 1848 & 1850 *65*
Peter Parley's Annual, 1848 *60*
Peter the Goatherd, *52*, *83*
Picture Pleasure Book, The, 1852 *75*, **76**
Photograph Album, The, *34*, **35**, *78*
Photographic Primer, The, 1854 *33*, *81*
Photographic Studies, 1853 *78*
Photographic Tour among the Abbeys of
 Yorkshire, A, 1856 *82*
Photographic Views of the Progress of the
 Crystal Palace, Sydenham, 1855 *33*, *78*,
 81
Photographs of Twelve Drawings by Birket
 Foster, 1866 *89*
Picture-Book of Merry Tales, A, *83*
Pilgrim's Progress, The Story of the, 1858
 83
Playmate, The, 1847-48 *18*, **24**, *62*, **63**, *78*
Playmate, The, Second Series, 1849 *62*
PLEASURE BOOKS FOR CHILDREN, *69*
PLEASURE-BOOKS FOR YOUNG
 CHILDREN, 1851 *69*
Pleasures of Hope, The, 1855 *36*, *80*
Pleasures of the Country, The, 1851 *70*
Poe, Edgar Allan, The Poetical Works of,
 1853 *76*
 1858 *83*
Poems, & Songs by Robert Burns, 1858 *82*
Poems of Oliver Goldsmith, The, 1859 *36*,
 38, *84*
Poetic Prism, The, 1848 *65*
Poetical Works of Edgar Allan Poe, The,
 1853 *76*
 1858 *83*
Poetical Works of Oliver Goldsmith, The,
 1851 *71*, *84*

Poetical Works of Thomas Gray, The, 1859
 84
Poetry Book for Children, A, 1854 *79*
Poetry Book for Schools, A, 1861 *85*
Poetry of Nature, The, 1861 *85*
Poetry of the Year, 1853 *34*, *36*, *78*, *82*
Poets of the Elizabethan Age, The, 1862
 38, *42*, *86*
Poets of the West, The, 1859 *84*
Poets of the Woods, The, 1853 *34*,
 36, **37**, *78*, *82*
Poets' Wit and Humour, 1861 *85*
Poor Margaret, 1855 *81*
Popular Faery Tales, ad. *56*
Portraits of the Members and Associates of
 the Society of Painters in Watercolours,
 1864 *89*
Practice of Photography, The, 1853 *33*, *78*
Pretty Poll, 1854 *79*
Puck's Reports to Oberon, 1844 *49*, *50*
Puss in Boots *52*, *83*, *89*

Raffaelle Gallery, The, 1871 *91*
Rambles in the Rhine Provinces, 1868 *90*
Red Field, or a Visit to the Country, 1858 *83*
Renaissance of Art in Italy, The, 1883 *91*
Reynard the Fox (Home Treasury), 1843
 48, *50*
 ad. *56*
Reynard the Fox, 1845 *18n*, *58*
Rhymes and Roundelayes in Praise of a
 Country Life, 1857 **40**, *82*
 ad. *56*
Richmonds' Tour in Europe, The, c. 1856
 82
Rime of the Ancient Mariner, The, 1857 *80*
Rip Van Winkle, 1850 *70*
Robin Goodfellow, 1845 **6**, *52*, *83*
 ad. *53*
Robin Hood and his Merry Forresters, 1841
 1, *47*
 ad. *56*
Robinson Crusoe, see Life and Adventures,
 etc
Romantic Story of the Princess Rosetta, The,
 1845 *52*, *83*
 ad. *53*
Rosebud, the Sleeping Beauty, 1845 *49*
 ad. *56*
Royal Alphabet of Kings & Queens, The,
 1841 *47*
Royal House of Tudor, The, 1866 *89*

Sabbath Bells Chimed by the Poets, 1856
 34, *36*, *78*, *82*
Series of Photographic Pictures, A, *78*
Series of Photographic Pictures of Welsh
 Scenery, A, *78*
Shakespeare's Songs & Sonnets, 1862, **3**
 38, *80*, *86*
Siegfield the Dragon Slayer, 1848 see
 Heroic Life, etc
Sir Guy of Warwick, *50*, *52*, *83*
 ad. *53*
Sisters, The, 1844 *49*, *50*, *83*
 ad. *56*
Sixty Etchings of Reynard the Fox, 1845
 18n, *58*
Sleeping Beauty in the Wood, 1845 *49*, *83*
 ad. *55*
Songs for the Little Ones at Home, 1860
 42, *85*
Songs, Madrigals and Sonnets, 1849 *18*,
 22, **23**, *26*, *67*
Songs of the Brave, The Soldier's Dream,
 etc, 1856 **39**, *80*
STANDARD STORY BOOKS,
 ad. *56*
Stories Told in Pictures, 1852 *72*
Story Book of Country Scenes, A,
 ad. *54*
Story Book of the Seasons, A; Spring;
 Summer; Autumn; Winter, 1845-48
 57-58
 ad. *54*
STORY BOOKS FOR HOLIDAY HOURS,
 1845 *58*
 ad. *54*
Story of the Pilgrim's Progress, The, 1858 *83*
Story of Jack and the Giants, The, 1851
 30, *70*
Story of the Three Bears, 1850 *30*
Sweet and Pleasant History of Patient
 Grissell, The, 1845 *52*, *83*

Tales from Denmark, 1847 *61*, **64**
Tales from Spenser's Faery Queen, 1846
 49, *50*
 ad. *56*
Tales of the Heroes of Greece (see Heroic
 Tales of Ancient Greece)
Tales of the Kings of England, 1840, 1841
 1, *47*
Terra-cotta Bricks, (see Box of Terra-Cotta
 Bricks) *22*
Tesselated Pastime, 1843 **ii**, *22*, *48*, *50*
Three Bears, The Story of the, 1850 *30*
Times, The *12n*, *45n*
Traditional Faery Tales ad. *56*
Traditional Nursery Songs of England, 1843
 6, *48*, *49*
 ad. *55*
Treasury of Pleasure Books for Young and
 Old, A, 1851 *69*, *71*, *75*
Treasury of Pleasure Books for Young
 Children, A, 1850 *30*, **31**, *69*, *71*, *75*
Treasury of Pleasure Books for Young
 People, A, 1850's *84*
True Tale of Robin Hood, A, 1843 *50*, *52*,
 83
Turner's Liber Studiorum, 1861 *86*
Twenty Views in Gloucestershire, 1854
 33, *81*
Two Centuries of Song, 1867 *42*, *89*
Two Doves and other tales, The, c. 1845
 58
 ad. *54*
Two Flies, a Moral Song, The, 1847 *61*
Two Talismans, The, 1846 *61*

Value of Money, and other Tales, The, 1846
 61
Veritable History of Whittington and his Cat,
 The, 1847 *49*, *50*, *83*
Vicar of Wakefield, The, 1855 *81*
Village Queen, The, 1852 *72*
Village Tales from Alsatia, 1848 *65*
Village Tales from the Black Forest, 1846
 61
Visits to the Zoological Gardens, c. 1856 *82*

Water Fairy and other tales, The, 1846 *61*
 ad. *54*
Water Lily, The, 1854 *79*
Whist-Player, The, 1856 *26*
Wild Spring Flowers, c. 1853 *77*
Wonder Castle, c. 1853 *77*
Wood-Engraving, A Brief History of, from
 its Invention, 1895 *45*, *91*
Wood-Nymph, The, 1870 *42*, *90*
Words of Truth and Wisdom, 1848 *18*, *66*
Works of Art in Pottery, Glass and Metal,
 91

REFERENCE BOOKS

Treatise on Wood Engraving, A, by J.
 Jackson, 2nd edn. 1861 *85*
Pre-raphaelite Diaries and Letters, by W.
 M. Rossetti, 1900 *70*
Fifty years of public work of Sir Henry Cole,
 K.C.B., vol. II, 1884 *49*, *52*
Birket Foster, by H. M. Cundall, 1906
 45n, *84*, *86*
Journal, by T. J. Cobden-Sanderson *45*
Life of J. Wolf, The, by A. H. Palmer,
 1895 *36*, *36n*
After Work, E. Marston, 1904 *34*
Reminiscences of Edmund Evans, The, 1967
 34, *84*
Lewis Carroll, Photographer, by H.
 Gernsheim, 1949 *33n*
History of Photography, The, by H. & A.
 Gernsheim, 1969 *33n*
Classic Fairy Tales, The, by Iona & Peter
 Opie, 1974 *30*
Victorian Book Design, 2nd edn, by Ruari
 McLean, 1972 *18*, *70*, *78*
English Print, The, by Basil Gray, 1937
 16n
History of the Christmas Card, The, by
 George Buday, 1954 *12*